In AI We Trust

In AI We Trust

Power, Illusion and Control of
Predictive Algorithms

Helga Nowotny

polity

First published in 2021 by Polity Press

Polity Press
65 Bridge Street
Cambridge CB2 1UR, UK

Polity Press
101 Station Landing
Suite 300
Medford, MA 02155, USA

ISBN-13: 978-1-5095-4881-1

A catalogue record for this book is available from the British Library.

Library of Congress Cataloging-in-Publication Data

Names: Nowotny, Helga, author.
Title: In AI we trust : power, illusion and control of predictive algorithms / Helga Nowotny.
Description: Medford, MA : Polity Press, 2021. | Includes bibliographical references and index. | Summary: "A highly original account of the nature of artificial intelligence and its implications for our future"-- Provided by publisher.
Identifiers: LCCN 2021003322 (print) | LCCN 2021003323 (ebook) | ISBN 9781509548811 (hardback) | ISBN 9781509548828 (epub)
Subjects: LCSH: Artificial intelligence.
Classification: LCC Q335 .N685 2021 (print) | LCC Q335 (ebook) | DDC 006.3--dc23
LC record available at https://lccn.loc.gov/2021003322
LC ebook record available at https://lccn.loc.gov/2021003323

Typeset in 10.5 on 12pt Sabon
by Fakenham Prepress Solutions, Fakenham, Norfolk NR21 8NL
Printed and bound in Great Britain by TJ Books Ltd, Padstow, Cornwall

For further information on Polity, visit our website:
politybooks.com

Contents

Acknowledgements

This book has been in the making for a long time, the strands of thought interrupted by travel and other obligations. There were several unsatisfactory starts and abandonments before what I was looking for came into focus. The actual process of writing benefited, perhaps paradoxically, from several COVID-19 lockdown periods, with the obligatory and, in my case, productive solitude they imposed. As an unprecedented event for everyone living in the twenty-first century, the pandemic also led me to reflect on its many unintended consequences, including some that have a direct bearing on the themes of this book.

My thanks go to all those who supported me in different ways on this long journey. Jean-Luc Lory offered me again a brief stay of hospitality at the Maison Suger in Paris, a wonderful place of calm in the centre of a buzzing metropolis. Another brief stay that I looked forward to, at the Wissenschaftskolleg zu Berlin, had to be cancelled due to abruptly imposed travel restrictions. Despite the missed occasion I was able to remain in fruitful exchange with Elena Esposito. I also want to thank Vittorio Loreto, of the Sony Lab Paris, for an especially inspiring conversation over dinner in Vienna. While the meal was frugal due to constraining circumstances, the discussion was rich. Stefan Thurner, who heads the Complexity Science Hub in Vienna, has been a continued source of inspiration and productive criticism. I received valuable feedback from him on parts of the manuscript, for which I am very grateful. Special thanks go to Michele Lamont at Harvard University for intellectual and moral support throughout the long stretch of the gestation period. Ever since we first talked about the book over a delicious lunch in New York City, Michele has responded to all my queries and provided me with continued encouragement.

When I approached John Thompson of Polity Press at the beginning of summer 2020 with an outline of the book, he responded without hesitation. Ever since, the process of getting the book published has unfolded in an ideal spirit of cooperation. I also want to thank the anonymous reviewers organized by Polity. One of them especially provided me with precious and concise feedback. I am also extremely lucky to have a splendid grand-daughter, Isabel Frey, who volunteered to be my personal editorial assistant. Isabel is not only a wonderful singer of Yiddish protest songs, but an engaged activist and researcher who represents the younger generation acting to change the world. Barbara Blatterer, my long-term personal assistant, took meticulous and efficient care in getting the manuscript through several versions to the final and decisive cuts, advising me also on the book's cover. To all of you, my heartfelt thanks.

Carlo Rizzuto acted again as my very special personal advisor, reading, commenting and encouraging me when I was doubtful whether I would ever finish. His unfailing humour provided me with the necessary distance from what I was doing. My thanks to him are as much for the ways in which he interrupted my writing as for everything else he gave me during this period.

Vienna and Bonassola, January 2021

Introduction: A Personal Journey into Digi-land

Origins: time and uncertainty; science, technology and society

This book is the outcome of a long personal and professional journey. It brings together two strands of my previous work while confronting the major societal transformations that humanity is undergoing right now: the ongoing processes of digitalization and our arrival in the epoch of the Anthropocene. Digitalization moves us towards a co-evolutionary trajectory of humans and machines. It is accompanied by unprecedented technological feats and the trust we put into Artificial Intelligence. But there are also concerns about continuing losses of privacy, what the future of work will be like, and the risks AI may pose for liberal democracies. This creates widespread feelings of ambivalence: we trust in AI as a bet on our future, but we also realize that there are reasons for distrust. We are learning to live with the digital devices we cheerfully interact with as though they were our new relatives, our digital others, while retaining a profound ambivalence towards them and the techno-corporate complex that produces them.

The process of digitalization and datafication coincides with the growing awareness of an environmental sustainability crisis. The impact of climate change and the dire state of the ecosystem upon which we depend for survival call for urgent action. But we are equally in thrall to or anxious about the digital technologies that are sweeping across our societies. Rarely, however, are these two major transformations – digitalization and the transition towards sustainability – thought together. Never before have we had the technological instruments and the scientific knowledge to see so far back into the past and ahead into the future, nor the techno-scientific capabilities for action. And yet, we feel the need to reconsider our existence in this uncanny present that marks a transition towards an unknown future that will be different from what has been promised to us in the past. This widespread feeling of anxiety has only been exacerbated by the COVID-19 pandemic, itself a major disruptive event with long-term consequences at a global scale.

My journey leading up to this book was long and full of surprises. My previous work on time, especially the structure and experience of social time, led me to inquire how our daily exposure to and interaction with AI and the digital devices that have become our intimate companions alter our experience of time once again. How does the confrontation with geological timescales, long-term atmospheric processes or the half-life of the dissolution of microplastic and toxic waste affect the temporalities of our daily lives? How does AI impinge on the temporal dimension of our relationship with each other? Are we witnessing the emergence of something we can call 'digital time' that has now intruded into the familiar nested temporal hierarchy of physical, biological and social times? If so, how do we negotiate and coordinate these different kinds of time as our lives unfold?

The other strand of my previous work, on uncertainty, directed my inquiry towards ways of coping with and managing old and new uncertainties with the help of the powerful computational tools that bring the future closer into the present. These tools allow glimpses into the dynamics of complex systems and, in principle, enable us to identify the tipping points at which systems transition and change the state they are in. Tipping points mark further transformation, including the possibility of collapse. As science begins to understand complex systems, how can this knowledge be harnessed to counteract the risks we face and strengthen the resilience of social networks?

Not surprisingly, I encountered several hurdles on my way, but I also realized that my previous long-standing interest in the study of time and the cunning of uncertainty – which, I argued, we should embrace – allowed me to connect aspects of my personal experience and biographical incidents with empirical studies and scientific findings. Such personal links, however, no longer seemed available when confronting the likely consequences of climate change, loss of biodiversity and the acidification of oceans, or issues like the future of work when digitalization begins to affect middle-class professionals. Like many others confronted with media images of disastrous wildfires, floods and rapidly melting arctic ice, I could see that the stakes had become very high. I kept reading scientific reports that put quantitative estimates on the timelines when we would reach several of the possible tipping points in further environmental degradation, leading to the collapse of the ecosystem. And, again like many others, I felt exposed to the worries and hopes, the opportunities and likely downsides, connected with the ongoing digitalization.

Yet, despite all these observations and analyses, a gap remained between the global scale on which these processes unfolded and my personal life which, fortunately, continued without major perturbations. Even the local impacts were being played out either in far-away places or remained local in the sense that they were soon to be overtaken by other local events. Most of us are cognizant that these major societal transformations will have huge impacts and numerous unintended consequences; and yet, they remain on a level of abstraction that is so overwhelming it is difficult to grasp intellectually in all its complexity. The gap between knowing and acting, between personal insight and collective action, between thinking at the level of the individual and thinking institutions globally, appears to shield us from the immediate impact that these far-reaching changes will have.

Finally, it struck me that there exists an entry point that allows me to connect curiosity-driven and rigorous scientific inquiry with personal experience and intuition about what is at stake: the increasingly important role played by prediction, in particular by predictive algorithms and analytics. Prediction, obviously, is about the future, yet it reacts back on how we conceive the future in the present. When applied to complex systems, prediction faces the non-linearity of processes. In a non-linear system, changes in input are no longer proportional to changes in output. This is the reason why such systems appear as unpredictable or chaotic. Here we are: we want to expand the range of what can be reliably predicted, yet we also realize that complex systems defy the linearity that still underpins so much of our thinking, perhaps as a heritage of modernity.

The behaviour of complex systems is difficult for us to grasp and often appears counter-intuitive. It is exemplified by the famous butterfly effect, where the sensitive dependence on initial conditions can result in large differences at a later stage, as when the flapping of a butterfly's wings in the Amazon leads to a tornado making landfall in Texas. But such metaphors are not always at hand, and I began to wonder whether we are even able to think in non-linear ways. Predictions about the behaviour of dynamic complex systems often come in the garb of mathematical equations embedded in digital technologies. Simulation models do not speak directly to our senses. Their outcome and the options they produce need to be interpreted and explained. Since they are perceived as being scientifically objective, they are often not questioned any further. But then predictions assume the power of agency that we attribute to them. If blindly followed, the predictive power of algorithms turns into a self-fulfilling prophecy – a prediction becomes true simply because people believe in it and act accordingly.

So, I set out to bridge the divide between the personal, in this case the predictions we experience as being addressed to us as individuals, and the collective as represented by complex systems. We are familiar and at ease with messages and forms of communication at the inter-personal level, while, unless we adopt a professional and scientific stance, we experience everything connected with a system as an external, impersonal force that impinges on us. Might it not be, I wondered, that we are so easily persuaded to trust a predictive algorithm because it reaches us on a personal level, while we distrust the digital system, whatever we mean by it or associate with it, because it is perceived as impersonal?

In science, we speak about different levels, organized in hierarchical ways, with each level following its own rules or laws. In the social sciences, including economics, the gap persists in the form of a micro-level and macro-level divide. But none of the epistemological considerations that follow seemed to provide what I was looking for: a way of seeing across these divides, either by switching perspectives or, much more challenging, by trying to find a pluri-perspectival angle that would allow me access to both levels. I have therefore tried to find a way to combine the personal and the impersonal, the effect of predictive algorithms on us as individuals and the effects that digitalization has on us as societies.

Although most of this book was written before a new virus wreaked havoc around the globe, exacerbated by the uncoordinated and often irresponsible policy response that followed, it is still marked by the impact of the COVID-19 pandemic. Unexpectedly, the emergence of the coronavirus crisis revealed the limitations of predictions. A pandemic is one of those known unknowns that are expected to happen. It is known that more are likely to occur, but it is unknown when and where. In the case of the SARS-CoV-2 virus, the gap between the predictions and the lack of preparedness soon became obvious. We are ready to blindly follow the predictions algorithms deliver about what we will consume, our future behaviour and even our emotional state of mind. We believe what they tell us about our health risks and that we should change our lifestyles. They are used for police profiling, court sentencing and much more. And yet we were unprepared for a pandemic that had been long predicted. How could this have gone so wrong?

Thus the COVID-19 crisis, itself likely to turn from an emergency into a more chronic condition, strengthened my conviction that the key to understanding the changes we are living through is linked to what I call the paradox of prediction. When human behaviour, flexible and adaptive as it is, begins to conform to what the predictions foretell, we risk returning to a deterministic world, one in which the future has already been set. The paradox is poised at the dynamic but volatile interface between present and future: predictions are obviously about the future, but they act directly on how we behave in the present.

The predictive power of algorithms enables us to see further and to assess the various outcomes of emergent properties in complex systems obtained through simulation models. Backed by vast computational power, and trained on an enormous amount of data extracted from the natural and social world, we can observe predictive algorithms in action and analyse their impact. But the way we do this is paradoxical in itself: we crave to know the future, but largely ignore what predictions do to us. When do we believe them and which ones do we discard? The paradox stems from the incompatibility between an algorithmic function as an abstract mathematical equation, and a human belief which may or may not be strong enough to propel us to action.

Predictive algorithms have acquired a rare power that unfolds in several dimensions. We have come to rely on them in ways that include scientific predictions with their extensive range of applications, like improving weather forecasts or the numerous technological products designed to create new markets. They are based on techniques of predictive analytics that have resulted in a wide range of products and services, from the analysis of DNA samples to predict the risk of certain diseases, to applications in politics where the targeting of specific groups whose voting profile has been established through data trails has become a regular feature of campaigning. Predictions have become ubiquitous in our daily lives. We trade our personal data for the convenience, efficiency and cost-savings of the products we are offered in return by the large corporations. We feed their insatiable appetite for more data and entrust them with information about our most intimate feelings and behaviour. We seem to have embarked on an irreversible track of trusting them. Predictive analytics reigns supreme in financial markets where automated trading and fintech risk assessments were installed long ago. They are the backbone of the military's development of autonomous weapons, the actual deployment of which would be a nightmare scenario.

However, the COVID-19 pandemic has revealed that we are far less in control than we thought. This is not due to faulty algorithms or a lack of data, although the pandemic has revealed the extent of grossly underestimating the importance of access to quality data and its interoperability. There was no need for predictive algorithms to warn of future epidemics; epidemiological models and Bayesian statistical reasoning were sufficient. But the warnings went unheard. The gap between knowing and doing persists if people do not want to know or offer many reasons to justify their inaction. Thus, predictions must also always be seen in context. They can fall on fallow ground or lure us into following them blindly. Predictive analytics, although couched in the probabilities of our ignorance, comes as a digital package that we gladly receive, but rarely see a need to unpack. They appear as refined algorithmic products, produced by a system that appears impenetrable to most of us, and often jealously guarded by the large corporations that own them.

Thus, the observations made during my patchy journey began to converge on the power of prediction and especially the power exerted by predictive algorithms. This allowed me to ask questions such as 'how does Artificial Intelligence change our conception of the future and our experience of time?' I could return to my long-standing involvement with the study of social time, and in particular the concept of *Eigenzeit*, which was the subject of a book I wrote in the late 1980s. A few years ago I followed up with 'Eigenzeit. Revisited', in which I analysed the changes introduced through our interaction with digital media and devices that had by then become our daily companions (Nowotny 2017). New temporal relationships have emerged with those who are physically distant but digitally close, so that absence and presence as well as physical and digital location have converged in an altered experience of time.

Neither I nor others could have imagined the meaning that terms like physical and social distancing would acquire only a few years later. In the midst of the COVID-19 pandemic, I saw my earlier diagnosis about an extended present confirmed. My argument had been that the line separating the present from the future was dissolving as the dynamics of innovation, spearheaded by science and technology, opened up the present to the many new options that were becoming available. The present was being extended as novel technologies and their social selection and appropriation had to be accommodated. Much of what had seemed possible only in a far-away future now invaded the present. This altered the experience of time. The present was becoming both compressed and densified while extending into the immediate future (Nowotny 1989).

What I observe now is that the future has arrived. We are living not only in a digital age but in a digital time machine. A machine fuelled by predictive algorithms that produce the energy to thrust us beyond the future that has arrived into an unknown future that we desperately seek to unravel. Hence, we scramble to compile forecasts and engage in manifold foresight exercises, attempting to gain a measure of control over what appears otherwise uncontrollable because of its unpredictable complexity. Predictive algorithms and analytics offer us reassurance as they lay out the trajectories for future behaviour. We attribute agency to them and feel heartened by the messages they deliver on the predictions that concern us most. Such is our craving for certainty that even in cases when the forecast is negative, we feel relieved that we at least know what will happen. In offering such assurance, algorithmic predictions can help us to cope with uncertainty and, at least partly, give us back some control of the future.

My background in science and technology studies (STS) allowed me to bridge the gap between science and society and reach a better understanding of the frictions and mutual misunderstandings that beset this tenuous and tension-ridden relationship. STS opens up the possibility of observing how research is actually carried out in practice and allows us to analyse the social structures and processes that underpin how science works. The pandemic has merely added a new twist, albeit a largely unfortunate one. While at the beginning of the pandemic science took centre-stage, combined with the expectation that a vaccine could soon be developed and therapeutic cures were in the pipeline, science soon became mired in political opportunism. A nasty 'vaccine nationalism' arose, while science was sidestepped by COVID-19 deniers and conspiracy theories that began to flourish together with anti-vax and extreme-right political movements. After a brief and bright interlude, the interface between science, politics and the public became troubled again.

The pandemic offered an advanced testing ground, especially for the biomedical sciences, whose recourse to

Artificial Intelligence and the most recent digital technologies proved to be a great asset. It allowed them to sequence the genomes of the virus and its subsequent mutations in record time, with researchers sharing samples around the world and repurposing equipment in their labs to provide added test facilities. It enabled the COVID-19 High Performance Consortium, a public-private initiative with the big AI players and NASA on board, to aggregate the computing capability of the world's fastest and most advanced computers. With the help of Deep Learning methods it was possible to reduce the 1 billion molecules analysed for potential therapeutic value to less than a few thousand.

The response to the pandemic also brought a vastly increased role for data. The pressure was enormous to proceed as quickly as possible with whatever data was available, in order to feed it into the simulation models that data scientists, epidemiologists and mathematicians were using to make forecasts. The aim was to predict the various trajectories the pandemic could take, plotting the rise, fall or flattening of curves and analysing the implications for different population groups, healthcare infrastructure, supply chains and the expected socio-economic collateral damage. Yet, despite the important and visible role given to data throughout the COVID-19 pandemic, no quick quantitative data-fix emerged that would provide a solid basis for the measures to be taken. If the data quality is poor or the right kind of data does not exist, a supposed asset quickly turns into garbage that contaminates simulation models and radically reduces their usefulness for society.

To some extent, the COVID-19 crisis has overshadowed the ongoing discussion about innovation and how scientific findings are transferred into society. It is therefore appropriate to recall the work of STS scholars who have extensively analysed the social shaping of technologies. Their findings show that technologies are always selectively taken up. They are gendered. They are appropriated and translated into products around which new markets emerge that give another boost to global capitalism. The benefits of technological innovation are never equally distributed, and already existing social inequalities are deepened through accelerated technological change. But it is never technology alone that acts as an external force bringing about social change. Rather, technologies and technological change are the products and the outcome of societal, cultural and economic preconditions and result from many co-productive processes.

Seen from an STS perspective, what is claimed to be entirely novel and unique calls for contextualization in historical and comparative terms. The current transformation can be compared to previous techno-economic paradigm shifts that also had profound impacts on society. In the age of modernity, progress was conceived as being linear and one-directional. Spearheaded and upheld by the techno-sciences, the belief was that continued economic growth would assure a brighter and better future. It came with the promise of being in control, manifest in the overconfidence that was projected into planning. This belief in progress has, however, been on the wane for some time, and more recently many events and developments have injected new doubts. The destruction of the natural environment on a global scale confronts all of us with an 'inconvenient truth', reconfirmed by the Fridays for Future movement that has galvanized the younger generation. In addition, the pandemic has demonstrated the helplessness of many governments and the cynicism of their responses, while coping with the long-term consequences will require a change in direction.

The remarkable speed of recent advances in AI and its convergence with the sustainability crisis invites the question: What is different this time? We are already becoming conscious of the limitations of our spatial habitat, and face multiple challenges when it comes to using the available resources in a sustainable manner. These range from managing the transition to clean energy, to maintaining biodiversity and making cities more liveable, to drastically curbing plastic pollution and managing the increasing amount of waste. No wonder there is a growing concern that the control we can exert will be further diminished. The machines we have created are expected to take over many jobs currently performed by humans, but our capacity for control will shrink even further because these machines will monitor and limit our actions and possibilities. For these reasons much wisdom will be needed to better understand how AI affects and limits human agency.

I soon realized that I had touched only the surface of deeper transformational processes that we will have to think about together. The future will be dominated by digital technologies while we simultaneously face a sustainability crisis, and both of these transitions are linked with changes in the temporal structures and regimes that shape our lives and society. Digital technologies bring the future into the present, while the sustainability crisis confronts us with the past and challenges us to develop new capabilities for the future. Whatever solutions we come up with must integrate the human dimension and our altered relationship to the natural and technologically transformed environment. These were some of the underlying questions that kept me going, humming quietly but persistently in the background while I continued my search. My journey took me to a number of international meetings, workshops and conferences where some of these issues were discussed. For example, there were meetings on how to protect rights to privacy, which received special legal status in Europe through the General Data Protection Regulation (GDPR). Europe is perceived to play only a side role in the geopolitical competition between the two AI superpowers, the United States and China, a competition sometimes referred to as the digital arms race for supremacy in the twenty-first century, and which has recently been rekindled in alarming ways. Many Europeans take solace in the fact that they at least have a regulatory system to protect them, even if they acknowledge that neither the GDPR nor other forms of vigilance against intrusion by the large transnational corporations are sufficient in practice.

Other items on the agenda of discussion fora about digitalization were concerned with the risks arising from the ongoing processes of automation. Foremost was the burning issue of the future of work and the potential risks that digital-ization entails for liberal democracies. It seemed to me that the fear that more jobs would be lost than could be created in time was being felt much more strongly in the United States than in Europe, partly due to still-existing European welfare provisions and partly because digitalization had not yet visibly hit professionals and the middle class. The threats to liberal democracies became more apparent when populist, nationalist and xenophobic waves swept across many countries. They were nurtured by sinister phenomena such as 'fake news' and Trojan horses, with unknown hackers and presumed foreign secret services engaged in micro-targeting specific groups with their made-up messages. More generally, they appeared intent on undermining existing democratic institutions while supporting political leaders with authoritarian tendencies. Digital technologies and social media were being appropriated as the means to erode democratic principles and the rule of law, while the internet, it seemed, had turned into an unrestrained and unregulated space for the diffusion of hate and contempt.

My regular visits to Singapore provided a different angle on how societies might embrace digitalization, and a unique opportunity to observe a digitally and economically advanced country in action. I gathered insights into Singapore's much-vaunted educational system, and observed the reliance of the bureaucracy on digital technologies but also its high standards of efficiency and maintenance of equally high levels of trust in government. What impressed me most, however, was the country's delicate and always precarious balance between a widely shared sense of its vulnerability – small, without natural resources and surrounded by large and powerful neighbours – and the equally widely shared determination to be well prepared for the future. Here was a country that perceived itself as still being a young nation, drawing much of its energy from the remarkable economic wealth and social well-being it has achieved. This energy now had to be channelled into a future it was determined to shape. Nowhere else did I encounter so many debates, workshops, reports and policy measures focused on a future that, despite remaining uncertain, was to be deliberated and carefully planned for, taking in the many contingencies that would arise. Obviously, it would be a digital future. The necessary digital skills were to be cultivated and all available digital tools put to practical use.

More insights and observations came from attending international gatherings on the future of Artificial Intelligence. In my previous role as President of the European Research Council (ERC), I participated in various World Economic Forum meetings. The WEF wants to be seen as keenly engaged in digital future building. At the meetings I attended, well-known figures from the world of technology and business mingled with academics and corporate researchers working at the forefront of AI. It was obvious that excitement about the opportunities offered by digital technologies had to be weighed against their possible risks if governments and the corporate world wanted to avert a backlash from citizens concerned about the pace of technological change. The many uncertainties regarding how this would be played out were recognized, but the solutions offered were few.

Other meetings in which I participated had the explicit aim of involving the general public in a discussion about the future of AI, such as the Nobel Week Dialogue 2015 in Gothenburg, or the Falling Walls Circle in Berlin in 2018. There were also visits to IT and robotics labs and workshops tasked with setting up various kinds of digital strategies. I gained much from ongoing discussions with colleagues at the Vienna Complexity Science Hub and members of their international network, allowing me glimpses into complexity science. By chance, I stumbled into an eye-opening conference on digital humanism, a trend that is gradually expanding to become a movement.

Scattered and inconclusive as these conversations mostly were, they nevertheless projected the image of a dynamic field rapidly moving forward. The main protagonists were eager to portray their work as incorporating their responsibility of moving towards a 'beneficial AI' or similar initiatives. There was a notable impatience to demonstrate that AI researchers and promoters were aware of the risks involved, but the line between sincere concern and the insincere attempts of large corporations to claim 'ethics ownership' was often blurred as well. Human intelligence might indeed one day be outwitted by AI, but the discussants seldom dwelt on the difference between the two. Instead, they offered reassurances that the risks could be managed. Occasionally, the topic of human stupidity and the role played by ignorance were touched upon as well. And at times, a fascination with the 'sweetness of technology' shimmered through, similar to that J. Robert Oppenheimer described when he spoke about his infatuation with the atomic bomb.

At one of the many conferences I attended on the future of AI, the organizers had decided to use an algorithm in order to maximize diversity within each group. The AI was also tasked to come up with four different haikus, one for each group. (Incidentally, the first time an AI succeeded in accomplishing such a 'creative' task was back in the 1960s.) The conference was a success and the discussions within each 'haiku group' were rewarding, but somehow I felt dissatisfied with the haiku the AI had produced for my group. So, on the plane on my way back I decided to write one myself – my first ever. With beginner's luck the last line of my haiku read 'future needs wisdom'.

A haiku is said to be about capturing a fleeting moment, a transient impression or an ephemeral sensation. My impressions were obviously connected to the theme of the conference, the future of AI. 'Future needs wisdom' – the phrase stuck with me. Which future was I so concerned about? Would it be dominated by predictive algorithms? And if so, how would this change human behaviour and our institutions? What could I do to bring some wisdom into the future? What I have learned on my journey in digi-land is to listen carefully to the dissonances and overtones and to plumb the nuances and halftones; to spot the ambiguities and ambivalences in our approaches to the problems we face, and to hone the ability to glide between our selective memories of the past, a present that overwhelms us and a future that remains uncertain, but open.

The maze and the labyrinth

None of these encounters and discussions prepared me for the surprise I got when I began to scan the available literature more systematically. There is a lot of it out there already, and a never-ending stream of updates that keep coming in. I concluded that much of it must have been written in haste, as if trying to catch up with the speed of actual develop-ments. Sometimes it felt like being on an involuntary binge, overloaded with superfluous information while feeling intel-lectually undernourished. Most striking was the fact that the vast majority of books in this area espouse either an optimistic, techno-enthusiastic view or a dystopian one. They are often based on speculations or simply describe to a lay audience what AI nerds are up to and how digital technologies will change people's lives. I came away with a profound dissatisfaction about how issues and topics that I considered important were being treated: the approach was largely short-term and ahistorical, superficial and mostly speculative, often espousing a narrow disciplinary perspective, unable to connect technological developments with societal processes in a meaningful way, and occasionally arrogant in dismissing 'the social' or misreading it as a mere appendix to 'the technological'.

Plenty of books on AI and digitalization continue to flood the market. Most of the literature is written in an enthusiastic, technology-friendly voice, but there is also a sharp focus on the dark side of digital technologies. The former either provide a broad overview of the latest developments in AI and their economic benefits, or showcase some recently added features that are intended to alleviate fears that the machines will soon take over. The social impact of AI is acknowledged, as is the desirability of cross-disciplinary dialogue. A nod towards ethical considerations has by now become obligatory, but other problems are sidestepped and expected to be dealt with elsewhere. Only rarely, for instance, do we hear about topics like digital social justice. Finding my way through the copious literature on AI felt at times like moving through a maze, a deliberately confusing structure designed to prevent escape.

In this maze there are plenty of brightly lit pathways, their walls lined with the latest gadgetry, proudly displaying features designed to take the user into a virtual wonderland. The darker groves in the maze are filled with images and dire warnings of worse things to come, occasionally projecting a truly apocalyptic digital ending. Sci-fi occupies several specialized niches, often couched in an overload of technological imagination and an underexposed social side. In between there are a large number of mundane small pathways, some of which turn out to be blind alleys. One can also find useful advice on how to cope with the daily nitty-gritty annoyances caused by digital technologies or how to work around the system. Plenty of marketing pervades the maze, conveying a sense of short-lived excitement and a readiness to be pumped up again to deliver the next and higher dose of digital enhancement.

At times, I felt that I was no longer caught in a maze but in what had become a labyrinth. This was particularly the case when the themes of the books turned to 'singularity' and transhumanism, topics that can easily acquire cult status and are permeated by theories, fantasies and speculations that the human species will soon transcend its present cognitive and physical limitations. In contrast to a maze with its tangled and twisted features, dead ends and meandering pathways, a labyrinth is carefully designed to have a centre that can be reached by following a single, unicursal path. It is artfully, and often playfully, arranged around geometrical figures, such as a circle or a spiral. No wonder that labyrinths have inspired many writers and artists to play with these forms and with the meaning-laden concept of a journey. If the points of departure and arrival are the same, the journey between them is expected to have changed something during the course of it. Usually, this is the self. Hence the close association of the labyrinth with a higher state of awareness or spiritual enlightenment.

The labyrinth is an ancient cultic place, symbolizing a transformation, even if we know little about the rituals that were practised there. In the digital age, the imagined centre of the digital or computational labyrinth is the point where AI overtakes human intelligence, also called the singularity. At this point the human mind would be fused with an artificially created higher mind, and the frail and ageing human body could finally be left behind. The body and the material world are discarded as the newborn digital being is absorbed by the digital world or a higher digital order. Here we encounter an ancient fantasy, the recurring dream of immortality born from the desire to become like the gods, this time reimagined as the masters of the digital universe. I was struck by how closely the discussion of transcendental topics, like immortality or the search for the soul in technology, could combine with very technical matters and down-to-earth topics in informatics and computer science. I seemed that the maze could transform itself suddenly into a labyrinth, and vice versa.

In practice, however, gaps in communication prevail. Those

who worry about the potential risks that digital technologies pose for liberal democracies discover that experts working on the risks have little interest in democracy or much understanding of politics. Those writing on the future of work rarely speak to those engaged in the actual design of the automated systems that will either put people out of work or create new jobs. Many computer scientists and IT experts are clearly aware of the biases and other flaws in their products, and they deplore the constraints that come from being part of a larger technological system. But at heart they are convinced that the solutions to many of the problems besetting society will arise from technology. Meanwhile, humanists either retreat to their historical niche or act in defence of humanistic values. The often-stated goal of interdisciplinarity, it seems, is not yet much advanced in practice.

I came away from the maze largely feeling that it is an overrated marketplace where existing products are rapidly displaced by new ones selected primarily for their novelty value. Depending on the mood of potential buyers, utopian or dystopian visions would prevail, subject to market volatility. The labyrinth, of course, is a more intriguing and enchanting place where deep philosophical questions intersect with the wildest speculations. Here, at times, I felt like Ariadne, laying out the threads that would lead me out from the centre of the labyrinth. One of these threads is based on the idea of a digital humanism, a vision that human values and perspectives ought to be the starting point for the design of algorithms and AI systems that claim to serve humanity. It is based on the conviction that such an alternative is possible.

Another thread is interwoven with the sense of direction that takes its inspiration from a remarkable human discovery: the idea of the future as an open horizon, full of as yet unimaginable possibilities and inherently uncertain. The open horizon extends into the vast space of what is yet unknown, pulsating with the dynamics of what is possible. Human creativity is ready to explore it, with science and art at the forefront. It is this conception of the future which is at stake when predictive algorithms threaten to fill the present with their apparent certainty, and when human behaviour begins to conform to these predictions.

The larger frame of this book is set by a co-evolutionary

trajectory on which humankind has embarked together with the digital machines it has invented and deployed. Co-evolution means that a mutual interdependence is in the making, with flexible adaptations on both sides. Digital beings or entities like the robots created by us are mutating into our significant Others. We have no clue where this journey will lead or how it will end. However, in the long course of human evolution, it is possible that we have become something akin to a self-domesticating species that has learned to value cooperation and, at least to some extent, decrease its potential for aggression. That capacity for cooperation could now extend to digital machines. We have already reached the point of starting to believe that the algorithm knows us better than we know ourselves. It then comes to be seen as a new authority to guide the self, one that knows what is good for us and what the future holds.

The road ahead: how to live forward and understand life backwards

Scientific predictions are considered the hallmark of modern science. Notably physics advances by inventing new theoretical concepts and the instruments to test predictions derived from them. The computational revolution that began in the middle of the last century has been boosted by the vastly increased computational power and Deep Learning methods that took off in the twenty-first century. Together with access to an unprecedented and still growing amount of data, these developments have extended the power of predictions and their applicability across an enormous range of natural and social phenomena. Scientific predictions are no longer confined to science.

Ever since, predictive analytics has become highly profitable for the economy and pervaded the entire social fabric. The operation of algorithms underlies the functioning of technological products that have disrupted business models and created new markets. Harnessed by the marketing and advertisement industry, instrumentalized by politicians seeking to maximize votes, and quickly adopted by the shadowy world of secret services, hackers and fraudsters exploiting the anonymity of the internet, the use of predictive analytics has convinced consumers, voters and health-conscious citizens that these powerful digital instruments are there to serve our needs and latent desires.

Much of their successful spread and eager adoption is due to the fact that the power of predictive algorithms is performative. An algorithm has the capability to make happen what it predicts when human behaviour follows the prediction. Performativity means that what is enacted, pronounced or performed can affect action, as shown in the pioneering work on the performativity of speech acts and non-verbal communication by J. L. Austin, Judith Butler and others. Another well-known social phenomenon is captured in the Thomas theorem – 'If men define situations as real, they are real in their consequences' – dating back to 1928 and later reformulated by Robert K. Merton in terms of self-fulfilling prophecy. The time has come to acknowledge what sociologists have long since known and apply it also to predictive algorithms.

The propensity of people to orient themselves in relation to what others do, especially in unexpected or threatening circumstances, enhances the power of predictive algorithms. It magnifies the illusion of being in control. But if the instrument gains the upper hand over understanding we lose the capacity for critical thinking. We end up trusting the automatic pilot while flying blindly in the fog. There are, however, situations in which it is crucial to deactivate the automatic pilot and exercise our own judgement as to what to do.

When visualizing the road ahead, I see a situation where we have created a highly efficient instrument that allows us to follow and foresee the evolving dynamics of a wide range of phenomena and activities, but where we largely fail to understand the causal mechanisms that underlie them. We rely increasingly on what predictive algorithms tell us, especially when institutions begin to align with their predictions, often unaware of the unintended consequences that will follow. We trust not only the performative power of predictive analytics but also that it knows which options to lay out for us, again without considering who has designed these options and how, or that there might be other options equally worth considering.

At the same time, distrust of AI creeps in and the concerns grow. Some of them, like the fears about surveillance or the future of work, are well known and widely discussed. Others are not so obvious. When self-fulfilling prophecies begin to proliferate, we risk returning to a deterministic worldview in which the future appears as predetermined and hence closed. The space vital to imagining what could be otherwise begins to shrink. The motivation as well as the ability to stretch the boundaries of imagination is curtailed. To rely purely on the efficiency of prediction obscures the need for understanding why and how. The risk is that everything we treasure about our culture and values will atrophy.

Moreover, in a world governed by predictive analytics there is neither a place nor any longer the need for account-ability. When political power becomes unaccountable to those over whom it is exercised, we risk the destruction of liberal democracy. Accountability rests on a basic understanding of cause and effect. In a democracy, this is framed in legal terms and is an integral part of democratically legitimated institutions. If this is no longer guaranteed, surveillance becomes ubiquitous. Big data gets even bigger and data is acquired without understanding or explanation. We become part of a fine-tuned and interconnected predictive system that is dynamically closed upon itself. The human ability to teach to others what we know and have experienced begins to resemble that of a machine that can teach itself and invent the rules. Machines have neither empathy nor a sense of responsibility. Only humans can be held accountable and only humans have the freedom to take on responsibility.

Luckily, we have not arrived at this point as yet. We can still ask: Do we really want to live in an entirely predictable world in which predictive analytics invades and guides our innermost thoughts and desires? This would mean renouncing the inherent uncertainty of the future and replacing it with the dangerous illusion of being in control. Or are we ready to acknowledge that a fully predictable world is never achievable? Then we would have to muster the courage to face the danger that a falsely perceived deterministic world implies. This book has been written as an argument against the illusion of a wholly predictable world and for the courage – and wisdom – needed to live with uncertainty.

Obviously, my journey does not end there. 'Life can only be understood backwards, but it must be lived forward.' This quotation from Søren Kierkegaard awaits an interpretation in relation to our movements between online and offline worlds, between the virtual self, the imagined self and the 'real' self. How does one live forward under these conditions, given their opportunities and constraints? The quotation implies a disjunction between Life as an abstraction that transcends the personal, and living as the conscious experience that fills every moment of our existence. With the stupendous knowledge we now have about Life in all its diversity, forms and levels, about its origins in the deep past and its continued evolution, is not now the moment to bring this knowledge to bear on how to live forward? The human species has overtaken biological evolution whose product we still are. Science and technology have enabled us to move forward at accelerating speed along the pathways of a cultural evolution that we are increasingly able to shape.

And yet, here we are, facing a global sustainability crisis with many dire consequences and mounting geopolitical tensions. As I write, we are in the grip of a pandemic, with others to follow if the natural habitats of animals that carry zoonotic viruses capable of spreading to humans continue to be eroded. The deficiencies of our institutions, created in previous centuries and designed to meet challenges different from our own, stare us in the face. The spectre of social unrest and polarized societies has returned, when what is needed is greater social coherence, equality and social justice if we are to escape our current predicament.

We have embarked on a journey to live forward with predictive algorithms letting us see further ahead. Luckily, we have become increasingly aware of how crucial access to quality data of the right kind is. We are wary about the further erosion of our privacy and recognize that the circulation of wilful lies and hate speech on social media poses a threat to liberal democracy. We put our trust in AI while we also distrust it. This ambivalence is likely to last, for however smart the algorithms we entrust with agency when living forward in the digital age may be, they do not go beyond finding correlations.

Even the most sophisticated neural networks modelled on a simplified version of the brain can only detect regularities and identify patterns based on data that comes from the past. No causal reasoning is involved, nor does an AI pretend that it is. How can we live forward if we fail to understand Life as it has evolved in the past? Some computer scientists, such as Judea Pearl and others, deplore the absence of any search for cause–effect relationships. 'Real intelligence', they argue, involves causal understanding. If AI is to reach such a stage it must be able to reason in a counterfactual way. It is not sufficient merely to fit a curve along an indicated timeline. The past must be opened up in order to understand a sentence like 'what would have happened if ...'. Human agency consists in what we do, but understanding what we did in the past in order to make predictions about the future must always involve the counterfactual that we could have acted differently. In transferring human agency to an AI we must ensure that it has the capacity to 'know' this distinction that is basic to human reasoning and understanding (Pearl and Mackenzie 2018).

The power of algorithms to churn out practical and measurable predictions that are useful in our daily lives – whether in the management of health systems, in automated financial trading, in making businesses more profitable or expanding the creative industries – is so great that we easily sidestep or even forget the importance of the link between understanding and prediction. But we must not yield to the convenience of efficiency and abandon the desire to understand, nor the curiosity and persistence that underpin it (Zurn and Shankar 2020).

Two different ways of thinking about how to advance have long existed. One line of thought traces its lineage to the ancient fascination with automata and, more generally, to the smooth functioning of the machines that have fuelled technological revolutions, with their automated production lines devoted to increasing efficiency and lowering costs. This is where all the promises of automation enter, couched in wild technological dreams and imaginaries. Deep Learning algorithms will continue to equip computers with a statistical 'understanding' of language and thus expand their 'reasoning' capacity. There is confidence among AI practitioners that work on ethical AI is progressing well. The tacit assumption is that the dark side of digital technologies and all the hitherto unresolved problems will also be sorted out by an ultimate problem-solving intelligence, a kind of far-sighted, benign Leviathan fit to manage our worries and steer us through the conflicts and challenges facing humanity in the twenty-first century.

The other line of thinking insists that theoretical understanding is necessary and urgent, not only for mathematicians and computational scientists, but also for developing tools to assess the performance and output quality of Deep Learning algorithms and to optimize their training. This requires the courage to approach the difficult questions of 'why' and 'how', and to acknowledge both the uses and the limitations of AI. Since algorithms have huge implications for humans it will be important to make them fair and to align them with human values. If we can confidently predict that algorithms will shape the future, the question as to which kinds of algorithms will do the shaping is currently still open (Wigderson 2019).

Understanding also includes the expectation that we can learn how things work. If an AI system claims to solve problems at least as well as a human, then there is no reason not to expect and demand transparency and accountability from it. In practice, we are far from receiving satisfactory answers as to how the inner representations of AI work in sufficient detail, let alone an answer to the question of cause and effect. The awareness begins to sink in that we are about to lose something connected to what makes us human, as difficult to pin down as it is. Maybe the time has come to admit that we are not in control of everything, to humbly concede that our tenuous and risky journey of co-evolution with the machines we have built will be more fecund if we renew our attempt to understand our shared humanity and how we might live together better. We have to continue our exploration of living forward while trying to understand Life backwards and linking the two. Prediction will then no longer only map the trajectories of living forward for us, but will become an integral part of understanding *how* to live forward. Rather than foretelling *what* will happen, it will help us understand *why* things happen.

After all, what makes us human is our unique ability to ask the question: *Why do things happen – why and how?*

1

Life in the Digital Time Machine

Between the mud and the sky: the birth of the digital Anthropocene

Beginnings are always difficult to pin down, especially when several strands come together. Human-induced global warming has convinced us that we have entered a new geological epoch, the Anthropocene. The designation remains unofficial as long as the required evidence for a 'Golden Spike' has not been approved by the official gatekeeper of geological time, the International Union of the Geological Sciences. The origins and scientific definition of the Anthropocene may still be unclear, but it is a juncture characterized by the entanglement of human activities on the human timescale with other temporalities. These include evolutionary timescales with different rates of species going extinct and new forms of life emerging, some of them created by us. We find ourselves confronted with the complexity of ecological timescales and ultimately with cosmic timescales subject to the laws of a still expanding universe. The Anthropocene thus contains a multitude of temporalities that reveal traces of the past, while pointing towards a future that harbours a vast space of possibilities. As I will argue, these temporalities also include digital time.

The Anthropocene leads us to reconsider our existence in a present that many find profoundly disconcerting. But its fuzzy origins reveal more than the ruins of a past that haunts us, and offer more than a constant reminder of the over-exploitation of natural resources that has led to dire warnings about future environmental collapse. The beginnings take us back to the radioactive traces left in the rocks at US nuclear test sites. First carried out overground in the 1940s, with massive radioactive fall-out, the tests were later moved underground. Now they offer evidence for the 'Golden Spike', the stratigraphic traces required to mark a new geological epoch. These developments led to the twin birth of nuclear power and the power that computation would assume in the digital age. They were enacted by the initial explosion of the atomic bomb and the proliferation of nuclear weapons that followed. The digital part proliferated as well. It has become ubiquitous and is now associated with with Artificial Intelligence, a term coined by Norbert Wiener in the mid-1950s.

George Dyson grew up on the premises of the Institute of Advanced Study (IAS), in Princeton, New Jersey, and has retraced what he calls the birth of the digital universe in the form of an origin story: 'There are two kinds of creation myths: those where life arises out of the mud, and those where life falls from the sky. In this creation myth, computers arose from the mud, and code fell from the sky' (Dyson 2012: ix). Origin stories are not easily translatable into precisely dated historical contexts. Should we identify the birth of the digital age with the brilliant insights of a young mathematician who in 1936 published his paper 'On Computational Numbers', introducing the formal hypothetical devices that would become known as Turing machines? Already in the seventeenth century Leibniz was firmly convinced that everything was calculable based on the dual principle of 0 and 1, and built a machine running on this principle in 1685. Others, like Charles Babbage and Ada Lovelace in the nineteenth century, were to follow. But it was in the mid-1940s at the IAS that a small group of physicists, mathematicians, biologists and engineers designed, built and programmed an electronic digital computer. Alan Turing's originality had provided the mathematical-logical spark that established the decisive link between code, now called software, and the arduous work of building physical machines, the hardware, that would run at electronic speed (Dyson 2012). It was this combination that led to the birth of the digital universe and, with it, of digital time.

But there is more to the origin story. The moment the digital universe and, implicitly, digital time were created eclipses the presence of human beings. Between the mud and the sky, humans occupy a middle ground. Humans built and deployed the atomic bomb. The scientific research and engineering efforts at the IAS in the 1940s were intimately connected with this historical moment of immense significance. The development of the bomb in Los Alamos was part of the wartime effort against Nazi Germany and its allies. It turned out to be instrumental for the advancement of electronic digital computers. Fission reactions had to be simulated accurately, which required computational aids to replace the work done manually by human 'computers'. When the first computational problem was run in 1945 on the newly designed and built Electronic Numerical Integrator and Computer, ENIAC, it was for the development of the hydrogen bomb.

Half a century later, the late Nobel laureate Paul Crutzen's suggestion of naming the onset of a new geological epoch 'the Anthropocene' became widely adopted. Long-term observations, measurements and modelling show the lasting impact that humans exert on the earth system, including its atmosphere and climate. The concept of the Anthropocene registers the dire state of the planet and represents a call for urgent action. On the middle ground between mud and sky, humans struggle to make their precarious living torn between the two. Having over-exploited the natural environment we now feel threatened by the consequences, which come with increasing frequency in the form of floods and droughts, melting arctic ice and the massive loss of biodiversity. Undeterred, we continue to reach out to the skies. Emboldened by the technologies at our disposal, we venture into outer space, exploring potential escape routes for the time when life on planet earth will become unliveable, all the while struggling to cope with the challenges we face right now.

Since 1947 the members of the Bulletin of Atomic Scientists, an association founded after the war by physicists involved in building the bomb, have released an annual report that is represented in the form of the Doomsday Clock, as a metaphor for how close humanity has moved towards a global catastrophe as a result of unchecked scientific and technological advances. Every year, the finger on the clock moves dangerously closer to midnight. It is no longer only the possibility of humanity annihilating itself that keeps the finger moving. Climate change and the collapse of vital parts of the earth system are now at the forefront of the impending catastrophe, while the global population continues to rise. For the first time ever, the metaphorical clock's time has fallen below the two-minute limit. Since 27 January 2021, the day of the last official announcement, it stands at 100 seconds before midnight.

Today, digital technologies are rapidly transforming our economies and societies. They are officially hailed by the European Union, among others, as the engine for a programmatic 'digital transition'. They have also led to a considerable expansion of the military arsenal to which a growing number of geopolitical players have access. The middle ground between mud and the sky continues to be contested over spheres of geopolitical influence, be it claims for uninhabited islands in the Pacific, for jungle patches occupied by insurgents who refuse to recognize a central government, or over access to resources in outer space. Even if the nuclear threat has lost the imminence it had during the Cold War, it has not vanished. It has merely receded into the background due to proliferation and dispersal. Meanwhile, humanity appears to be moving towards digitally advanced autonomous weapons systems, with drones already flying over enemy territory equipped with instrumentation that enables them to hit self-selected targets with precision.

From around the 1950s, the space between mud and sky, the middle ground inhabited by us, underwent a remarkable convergence of two large-scale developments that had not previously been seen as connected. Changes in human activities started to correlate strongly with changes observed in the earth system. With the enormous surge in economic growth – closely linked to the rise of GDP, population growth and primary energy use – many associated human activities began to show a strong correlation with the key indicators of changes being observed in the earth system, including greenhouse gas emissions, deforestation, ocean acidification and a host of other indicators that are now displayed and continuously updated on a planetary dashboard. The phenomenon is known as the Great Acceleration. What began some seven decades ago shows no signs of slowing down (Steffen et al. 2015; McNeill and Engelke 2015).

This co-occurrence of environmental transformation and innovative human activity appears not to be confined to the Great Acceleration. Much further back in time similar patterns have been observed, although on a much smaller scale. Historical records based on data that stretches back thousands of years, patchy and deficient as they are, show striking correlations between periods of major climatic change and bursts of innovative human activities. One hypothesis is that a high variability of climate selected for humans who had the capacity to adapt, either by speciation, migration or developing new tools (Slezak 2015).

In 1955, John von Neumann, the mathematician and engineer at the forefront of building the first operable computer in Princeton, wrote a brief essay with the startling title: 'Can We Survive Technology?' He referred to the most radical advanced technologies of his day, the atomic bomb and nuclear energy, and argued that with them the acceleration of technological change met its natural limit – the size of the earth. During the industrial age technological progress had been extended geographically, but with the advances in nuclear technology the propensity to evolve towards ever larger-scale operations and to extend spatially was halted. As most timescales are fixed by human reaction times, habits and other factors, technological acceleration, von Neumann argued, was no longer possible due to the heightened risk of mutual nuclear destruction on the part of the two superpowers. This induced the instability that worried him most and which, so he argued, had ended the further spatial expansion of technology (von Neumann 1955).

With the benefit of hindsight, it is a remarkable irony that one of the pioneers of computerization and digitalization did not foresee the enormous impact the new digital technologies would have through the widespread and decentralized diffusion of computers; that a technology he had helped to create would overcome the innate temporal limitations of humans, and that digital technologies in the form of satellite communication would expand the scope and scale of monitoring and surveillance of the finite earth. The transformation of the world by digital technologies has led to a compression of time, while the spatial reach within such compressed timescales has expanded. However, von Neumann's diagnosis that the ultimate limit for technology is the finiteness of the earth still stands, at least for now (Fleurbaey et al. 2018).

While the official timekeepers continue to deliberate over whether the geological traces in rocks or lake sediments resulting from human activities are sufficient to recognize the Anthropocene, the enormous challenges that come with the sustainability crisis have been summarized by the United Nations in seventeen 'sustainable development goals' (SDGs). The time pressure to ward off further deterioration, and possible collapse, is enormous. The intricate interconnections between the digital transition and the green transition are still not sufficiently recognized, but novel solutions may emerge by adopting more systemic, holistic and integrative approaches to bring them together (Renn 2020). This will require taking into account evolutionary perspectives and their timescales as well. The manifold interactions of the earth system with our activities are crucial for life on the planet – the only environmental niche the human species has for a population that will soon reach 8 billion inhabitants.

To find viable solutions without continuing unsustainable resource exploitation will require some kind of symbiotic relationship between the human-constructed environmental niche in which we live and the life-supporting systems of the Anthropocene. Many living organisms are known to alter their local environment and biologists argue that niche construction is an evolutionary process. The human species, just like octopuses or worms, has also been engaged in niche construction, adapting its behaviour to changes in the ecological conditions and in order to co-exist with other species (Laubichler and Renn 2015). In the epoch of the Anthropocene our niche construction is increasingly pursued with the help of digital technologies. It is nested in the natural environment but equipped with a vast computational infrastructure that enables continued monitoring through the collection of data, ranging from the flows of energy and people in megacities to equipping jellyfish with tiny sensors to measure ocean acidification. Satellites are sent into orbit, special vehicles explore deep ocean floors for mining purposes, and the ground beneath cities is mapped for urban expansion.

As no digital infrastructure can exist without first transforming matter, the extraction of minerals containing the forty or so chemical elements needed for the manufacture of smartphones, sensors and other digital devices, also continues. The amount of energy needed for cloud computing or for the operation of platforms and social networks, not to speak of blockchains, requires the much-discussed transition to sustainable energy to pick up speed. Digital technologies will have to play a central and responsible role if geo-anthropology, the emerging science of human–earth interaction, is to succeed (Rosol et al. 2018). An integrative perspective on the efforts to link the human microsphere with the planetary macrosphere is needed to make the middle ground sustainable.

We have entered a co-evolutionary path with the digital machines we have created. Although humans are the product of biological evolution, we have been able to overtake it by launching a cultural evolution largely based on science and technology. With digital technologies humans can do things that were previously unimaginable, even if we are not yet sure how best to use them or for which ultimate purposes. Yet, in many ways we remain tied to the origins and constraints of our biological evolution, even if we can now identify the genes inherited from the long and twisted lineage of our ancestors that have been retained or lost, and know which functions that were once beneficial for survival and adaptation have become obsolete or turned nefarious.

The heritage of evolution is evident when it comes to the experience of time. The biological arrow of time has been inscribed into us as into all living organisms. It leads from birth to death along the multiple pathways of ageing, even if our species has been uniquely successful in lengthening its lifespan. Technologies and the materials they are made of are also exposed to wear and tear. Things fall apart, erode and decay, and in this sense they age. Their remnants turn up as heaps of plastic, electronic and other waste that pollutes oceans and megacities. Digital code and other immaterial parts of technologies become obsolete even more quickly, overtaken by more recent innovations. But none of these processes are identical to those in living organisms. Life depends on the finely tuned interaction of the many rhythms and cycles that regulate everything from the inner dynamics of cells to the networks they form, from the developmental stages before birth to the amazing synchronization that takes place in the brain and guarantees the unique unity of mind and body.

It is against this backdrop that the impact of digitalization on the concept and experience of time stands out as marking another decisive shift. Social time constitutes a temporal order through which societies coordinate the activities of their members and their relation to themselves and to nature. It is a social construct that has to be constantly renegotiated along with the demands made by other temporalities and temporal orders. The industrial age brought with it clock time and assured the global dominance of the linear concept of time, replacing the cyclical time of preindustrial societies that connected the daily rhythms of human activities with those of Nature and the cosmos (Descola 2019). Industrialization and modernization linked technological acceleration with cultural and social acceleration. The linear time that underlies the modern age introduced a mixture of intense time pressure, exhaustion and the aspiration to a continuously improving future. It nurtured the wish for *Eigenzeit*, time of one's own (Nowotny 1989).

Now the linearity of our temporal experience has been broken. The Anthropocene obliges us to relate human timescales with ecological and planetary timescales. Social time has to accommodate the digital time that is built into the technologies that surround us. Thus the seemingly coherent, if individually exhausting, temporal framework that dominated modernity is no longer available in the digital Anthropocene. The configuration of past, present and future as a linear unfolding, the 'story of history itself', has crumbled. The future can no longer be perceived in terms of the telos built into it by the belief in progress, since the direct link between past accumulation and an ever-improving future no longer exists. The future has ceased to serve as 'the Eldorado of our hopes and wishes' (Jan Assmann). In fact, it can no longer be taken for granted, now that the linearity between yesterday and tomorrow has been broken. The relationship with the future has been upturned by the bewildering complexities of the Anthropocene, exposing the inadequacy of human timescales in relation to the required planning horizons. The entanglement of multiple temporalities pervading the Anthropocene generates a multitude of new, anthropocenic experiences of time in which the ruins and traces of the past undercut linear notions of time, while opening up the possibility of articulating alternative futures (Jorritsma 2020).

The anthropocenic experience of time reveals the different layers of the past, 'the sediments of time' (Koselleck 2018), with their continuities and ruptures that lead to the emergence of the new. But our experience of time is also challenged by the pervasive intrusion of digital time. Social scientists have extensively described the changes brought about by digital technologies in everyday life (Elliott 2019). The use of smartphones that fill every idle minute and waiting time is one of the many ways in which social and digital time are interacting and have to be renegotiated. Digital calendars are taking over shared social timekeeping. Intelligent digital assistants with their tracking capacities and behavioural algorithms are deployed to solve the problem of how best to organize one's time. Judy Wajcman has studied how designers and software engineers working on the automation of time management struggle with the question of how much agency to confer on the personal digital assistant acting on the user's behalf. When, at the end of a demonstration conference, a Google CEO assures the audience that 'we are working hard to give users back time', the question arises as to whose time this is, and how much control should be given to digitally managed time over the employee's discretion (Wajcman 2019).

But beyond these familiar transformations in our temporal relationships with each other and with ourselves – what is digital time? It is non-linear and permeates complex systems. When the digital universe was born, digital time entered the social world. It is the newcomer in the nested hierarchy of temporalities that encompasses temporal becoming, from the cosmic to the human and the 'noosphere', the realm of ideas, knowledge and consciousness (Fraser 1975). Born from a symbiosis of mathematical symbols and pieces of hardware made to run electronically and to affect how things are done, digital time pervades other temporalities and the spaces they occupy. But it does not claim a domain of its own. Viewed through the lens of digital time we can see the emergence of the new in real time, in our social time, enabling us to discover unseen patterns and finding hidden connections. This allows us to bring the past into the present and to glimpse possible futures ahead, thus subverting the modernist assumption of linear time that proceeds smoothly from past to present and future. We can now track a weather front, a traffic jam in the making or the collapse of a financial system in 'real time' as the dynamics underlying these processes unfold. The middle ground between mud and sky has turned into a tangle of different temporalities operating at different scales, mixing long-term and short-term cycles, while having to be integrated into our experience of social time.

The past reaches the present and the (visible) future has arrived

This temporal entanglement must be brought into some kind of temporal order in which digital time has to be mediated, negotiated and accommodated into social time. One way to achieve this is to think of living in a digital time machine. Unlike with the time machines from science fiction, there is no magic shuttle along the axis of linear chronological time where strange things happen, like becoming older than one's grandparents. The digital time machine operates in a different, far more radical way: we are not leaving the present at all, but we expand it. The digital time machine brings the past into the present where it affects the experience of the now. It makes us see the future in a different way. Digital time operates through algorithms, selecting and extrapolating from past data to foresee the future in the making. The digital time machine thrusts us forward into a future that fills us with a mixture of excitement and unease. But in contrast to the future imagined in science fiction, or enthusiastically celebrated by the futuristic movements that thrived during modernity, our excitement is muted. We feel unease because things might go wrong, because we are unsure how far we can trust the algorithms that tell us what will happen. It therefore matters all the more how we think about the future. Will we be able to think the future in a different way? Can we think it in a non-linear way?

Life in the digital time machine exposes a paradoxical situation. It allows us to see further ahead, but if we start to believe that the future ahead is the only possible future, we risk closing down other options. We may fall back into a linear temporality, forgetting that we have entered the digital Anthropocene, where non-linear connections prevail. Life in the digital time machine shifts our fundamental temporal bearings. The experience of past, present and future was never that rigid, and different cultures in different parts of the world and at different times have articulated alternative ways in connecting them. The past has sometimes been considered dangerous, so that special precaution had to be taken to guard against the ghosts of ancestors attempting to return to the company of the living. In medieval Europe, Christians were convinced that an afterlife in hell or in heaven awaited them, sharply segregating the idea of an immortal future after death from a moribund present. In the Western imagery, the future lies before us, but in other parts of the world one first has to go back in order to go forward, as the future may be hidden in the past.

Our present has become informationally and emotionally overloaded with virtual activities that fill what remains only a twenty-four-hour day. This leads to a peculiar, almost quantistic experience of the present in the digital time machine. We feel squeezed by the technology-induced compression of time into ever smaller units that demand ever greater efforts to assemble the disjunct parts into a meaningful whole. But the present is also experienced as extending towards the past as well as the future. Digital technologies enable access to the deep past. When the thawing Siberian permafrost revealed a well-preserved rhinoceros, for instance, its genetic lineage, how long ago it lived and how old it was when it presumably drowned, was determined by genetic sequencing.

Sophisticated scientific instrumentation lets us see the very big and the very small, the far and the near. We may zoom in and out, seeing the fine-grained details and the larger picture in motion, and manipulate objects and processes. We have become active participants in an interaction with a past that has been transferred into the present, be it through scientific research, simulations and games, in virtual space and in real time.

Nowhere is this more evident than when we look back into the deep past of the universe, enabled by the amazing advances in the techno-sciences. Nothing could be further away in time and space than events that occurred some unimaginable 1.3 billion years ago. Such an event was revealed in April 2019, when the first image of the shadow of a black hole was sent around the world. This followed decades of collaborative research efforts obtained from eight radio telescopes on five continents. The media echo was enormous. The image of a halo of hot gas and plasma – what in technical terms is called a black hole's event horizon – from an enormous elliptical galaxy some 53 million light years away, and holding the unimaginable mass of 6.5 billion suns, was brought into living rooms. Some saw in it what hell might look like and others the ultimate destiny of our sun. Even if this event had no direct impact on our everyday lives, it was felt as though it was happening right now, in the present.

If the past is another country, as the saying goes, many pathways lead there and even more interpretations of it are possible. The relationship with the past is always affected by the problems and concerns of the present. The further back the events, the easier it is to accommodate them. News about black holes or gravitational waves that reach us from outer space convey something of the vastness of the universe and its different timescales. They enter the collective imaginary as traces from the deep past. Taken together with the digital reach of orbiting spacecraft and telescopes, human exploration of outer space gathers pace. Every object in space that is spotted, recorded, identified and visualized is measured in time that equals distance expressed in the speed of light. We are in a process where a small but growing segment of the deep past might become our future, the next home for humanity.

A more uneasy relationship with the past stems from the analysis of ancient human DNA and the effect this has on our understanding of our common evolutionary history. The latest findings from paleogenomics and related disciplines, based on increasing computational power, newly available masses of data and new digital sequencing methods, often prove unsettling for the stereotypes, misinformation and scientifically untenable ideas about 'race' that are still rampant. The extraction of tiny bits of human DNA from otherwise contaminated complex fossilization processes opened the floodgates for more discoveries, beginning with the first Neanderthal genome, published in 2010. The Denisovan genome followed and subsequent ancient DNA discoveries continue to produce a torrent of findings about how the human species came to be what it is now.

Scientific findings from the past are not always welcome, especially when they contradict cherished narratives about who people think they are and where they think they come from. Evidence from ancient DNA shows the human evolutionary past to have involved a series of migrations and replacements, matings and extinctions. It renders visible the ghost lineages of early humans whose genes can still be detected but who went extinct. Vast population movements reaching back to the early diversification of human groups in Africa come to the fore, whose DNA can be matched with the languages spoken today. The past reaching the present holds more surprises and opens more questions that have to be accommodated in the present, complicating the ongoing search for identity that has become so central today.

Human history reveals itself as one of a continual mixing of populations over thousands of years, through conquest, migration, cultural assimilation and replacement. David Reich recounts a story of scientific collaboration with colleagues in India who had a vast collection of DNA samples representing the extraordinary human diversity in that country (Reich 2018). Their analysis showed that Indians today descend from a mixture of two highly divergent ancestral populations, one being 'West Eurasians' and the other distantly related to East Asians, but separated from them over thousands of years. When Reich discussed these initial findings with his Indian colleagues, their reaction almost scuttled the entire project. They hesitated to continue to be part of a study that suggested that a major West Eurasian immigration into India had occurred, arguing that they must be certain how the results could be reconciled with their own mitochondrial DNA findings.

It soon became clear, however, that the objections were rooted in the anticipation that the idea of a migration from outside India with such transformative effect would be politically explosive. Finally, a solution was found by changing the language. Today's Indians, the research findings would read, are the outcome of mixtures of two highly differentiated populations, renamed as 'Ancestral North Indians' and 'Ancestral South Indians', with everyone in India a mix, albeit in different proportions, of ancestry more closely related to one or the other.

The genetic study also revealed differences in status and social power between men and women. Males from populations with more power tended to pair with females from populations with less power. The ancient Indian caste system with its strict norms of endogamy clearly showed in the genetic data. It can cause population bottlenecks where rare disease-causing mutations carried by the founder individuals can dramatically increase in frequency. Building an ancient DNA Atlas of humanity is expected to tell us much about the frequency of biologically important mutations and the evolution of pathogens. Reich remains optimistic that the pursuit of truth for its own sake will overcome stereotypes, undercut prejudice and highlight the connections among people previously not known to be related. Society still has to catch up with the latest findings from science about our common past, but as long as politicians insist on 'alternative facts' there is still a long way to go.

In April 2019 there was another event that received huge global media coverage. The fire that broke out in Notre-Dame de Paris in the evening of 16 April completely destroyed the wooden parts of the cathedral's roof and much of the interior. The flames lasted for hours, but the spectacular fall of the spire seen by millions of viewers on TV took only a few minutes. Fortunately, the stone architecture with the vault in the middle and external support structures withstood the flames. This event, of course, was neither planned nor announced. It was entirely human-made and the timescale it covered reached back only a few centuries. It was an accident, albeit with devasting consequences, caused by negligence, human error and chains of miscommunication.

The event had worldwide resonance, mixing sadness and solidarity. Most striking was the wave of emphatic senti-ments that followed. For a brief historical moment, a church built 800 years ago, and occupying the centre of secular Paris in the twenty-first century, was reclaimed as belonging to humanity, regardless of religion, nationality, age or personal memories. Historians of art were quick to enter the mediatic scene, affirming that much of French and European history had been inscribed in the building. Technology quickly allowed the geographical origins of the oak beams to be traced back to a densely wooded area in Northern France that no longer exists. The cathedral had been exposed to the turbulence of the French Revolution and many other events of historical significance. In the ensuing days, other layers of different memories associated with Notre-Dame were laid bare. Events and commemorations of collective importance crisscrossed the personal memories of millions of people who had visited the cathedral at some point in their life. The past entered the present through the meanings attached to these individual as well as collective memories.

Memories from the past are subject to the passing of time. They cannot be preserved as they are and they gradually recede into the stored memory. It is tempting to look for similarities and deeper meanings when past events hit the public imagination, causing them to resonate above and beyond everyday affairs and mundane concerns. Whether it is the fire consuming Notre-Dame, the fiftieth anniversary of the Moon landing, the imaging of black holes or the tribute shown to outstanding individuals following their death, the common bond between such events is the attempt to transcend the present by reclaiming the past as belonging to it. It is a defiance in the face of forgetting and a mobilization of the collective will to build something which endures.

Perhaps this is what the world-renowned theoretical physicist Stephen Hawking had in mind when he once spoke about 'cathedral thinking' – far-reaching visions that are shared and that strengthen people's relations to each other.

The construction of cathedrals in various parts of Europe during the Middle Ages required the ability and commitment of a community of builders and their supporters to look ahead beyond the next generation. The builders were highly skilled craftsmen who moved freely throughout Europe in well-connected social networks among the different construction sites. Financing for cathedrals came from many sources, turning them into places that could be claimed by the local community. The time it took to build them was usually longer than predicted. Even a finished cathedral was never completely finished.

When George Dyson wrote the history of the Institute of Advanced Study in Princeton, he chose *Turing's Cathedral* as his title. This may not be a coincidence. Alan Turing was a 'cathedral thinker'. Cathedral thinking remains rare. If successful, it is one of the most impressive ways of bringing past and future together in the present. It humbly acknowledges that work performed in the present builds on what has been done in the past, while actively engaging with a future that transcends a lifetime and utilitarian gains. It is carried forward with the knowledge and intention that it will outlast the present, and is conducted in a team spirit that spans generations. Cathedral thinking and cathedral builders will be needed for our life in the digital time machine as well.

History is an essential part of meaning-making, even if history has no meaning of its own. The possibilities of learning from history remain limited, although looking back can provide salutary insights at a distance. We read into the past what preoccupies us in the present. Anxieties tend to return and the ghosts from the past are never far away. In a similar way, we approach the future in terms that are shaped by the present, by the hope or despair we feel, oscillating between confidence and hubris on one side and humility and openness on the other.

Thus, the COVID-19 pandemic that took us by surprise rendered visible old and new ruptures, while its longer-term consequences are still unfolding. It also pushed us into a digital way of life previously unimagined. Does this mean that digital time will also take over, severing further our ties to each other and to the natural environment, even if we know they are subject to change? Lockdowns and physical distancing, social isolation, home working and home schooling have given us a foretaste of how social relationships might evolve, while we are grateful for the digital technologies that kept us at least virtually connected. Life in the digital time machine enables new ways of seeing the past and the future together in the present. It offers us multiple perspectives and the possibility of adopting a more holistic approach, based on the ability to see things connected.

We were quick to familiarize ourselves with the future that has already arrived, at least its highly visible part. It greets us on the screens of smartphones and hovers in the sensors installed in surveillance cameras; in the drones that fly over us and the tiny robots that scan the insides of our bodies. This visible part of the future is everywhere recognizable in its pulsating connectivity, but it is equally dispersed and elusive. It arrives in highly visualized form, in images and videos, inviting us to share selfies and play games in an imaginary space. It is fast changing and difficult to grasp in the ephemeral forms it adopts and in its operational structures that remain hidden. Maybe this is why we keep adding to the already immense pool of visual data by taking and sharing millions of photographs. We laugh like children without understanding how our smile ends up instantaneously on the phone of a friend thousands of miles away. Once the photo album served to preserve the chronology of family life. Now nothing can be remembered if it has not been widely shared. It is as though we need to constantly reassure ourselves of our existence and fleeting identities.

More than a trillion photographs are now taken every year, up from 80 billion in 2000. Circulated on the internet they generate a booming business for the social media firms that store and distribute the soaring supply of pictures. The then only two-year-old Instagram was bought by Facebook in 2012 for $1 billion, and now has 1 billion active users with 95 million photographs and videos uploaded every day. Snapchat, Instagram's rival, rejected a $3 billion offer from Facebook in 2013, and did the same when Google offered $30 billion three years later (Earle 2020). Behind these takeover transactions lies the huge market of location, rental, restaurant, tourist and artistic sites. This market has to reach potential customers and let them see in advance what they will get, only to collect more data about their 'experience' afterwards.

The visible part of the future that has arrived enforces its visibility by spreading ever further a culture of the ephemeral image, all the while recognizing faces and transforming them into more data to train algorithms to recognize more faces even better. But there is also an invisible part in the operations performed by algorithms, search machines, bots and trolls. Most of it goes unnoticed. Much of it is hidden, often deliberately, not least by furtive hackers in unknown places working for shadowy companies or criminal bands. Little is known about who runs the bitcoin industry, but its stock price keeps rising on the markets. Another part of the invisible future is hidden in the corporate labs of the big corporations, where much of the cutting-edge research and development on algorithms is carried out. All these operations depend on the invisible technical infrastructures that supply the energy needed for broadband networks and electric grids.

The arrival of the future in the present adds to the informational and emotional overload that plagues us. Squeezed between the past and the future, the digital present is compressed and densified. Paradoxically, the future that is already here has dimmed the appetite for a future that once seemed highly desirable. The sci-fi writer William Gibson once famously said that the future is already here, only unevenly distributed. Recently, he revised this statement based on the observation that if too much of the future has to be digested in the present, it leads to wariness and even future-fatigue. During the twentieth century the future was a cult, if not a religion, but now interest in it is waning. Previously evoked with techno-enthusiasm and curiosity, interest in the twenty-second century has almost completely disappeared today (Spicer 2020).

The future that has been envisioned is never the future that actually arrives. We tend to overestimate the technologies that are in full view and have settled in our imagination, but are woefully incapable of imagining their social consequences. This is why reading predictions about technological developments that look exciting at the time can make for distressing reading fifty years later. This also holds for

Artificial Intelligence. As Jill Lepore observes, predictive algorithms

> start out as historians: they study historical patterns to detect patterns. Then they become prophets: they devise mathematical formulas that explain the pattern, test the formulas against historical data withheld for the purpose, and use the formulas to make predictions about the future. That's why Amazon, Google, Facebook, and everyone else are collecting your data to feed to their algorithms: they want to turn your past into your future. (Lepore 2018)

Will we let them do so? Life in the digital time machine brings the past into the present and integrates the visible part of the future that has arrived. It leads to an overload of the present, but also entails new ways of seeing things together. A past that is selectively remembered to instil continuity with the present is then no longer to be taken for granted, nor a future tied to the present as the accumulated outcome. The paradox of algorithmic predictions is that we are free to imagine alternative futures – unless we believe that those predictions are the only future we have.

The open horizon of the future

The desire to know the future is as old as humanity. All cultures are known to have practised some kind of divination to reveal what lies ahead. Traces of this are to be found everywhere, adapted to local means and circumstances. These practices share the assumption that destinies have been predetermined and are known to the gods, but human beings remain in ignorance. Therefore, trained practitioners and intermediaries with special skills are needed to interpret the signs transmitted from the otherworldly realm. In some parts of ancient China, for instance, divinatory experts held the shoulder blades of sheep or the shells of tortoises above fire to induce cracks in the bones that would then be interpreted by them. Chinese oracle bones are now thought to display what might have been the origins of the Chinese script. Inadvertently, the desire to know the future may have led to a technique of preserving the past for the future – the invention of writing.

Elena Esposito has drawn attention to some fascinating similarities between ancient practices of divination and the algorithms used today to make predictions about future human behaviour (Esposito 2021). Ever since the rise of statistics, modern societies have used them not only for administrative purposes but also as an efficient way to manage uncertainty. Based on large numbers, probability calculus and other statistical tools, patterns detected in the past were extrapolated for the future (Hacking 1975; Porter 1995). Contrasting the 'governance by numbers' based on statistics with the predictive algorithms that feed on big data today, Esposito notices a return to divinatory practices. Predictive algorithms do not address averages and general trends in the population, as administrative statistics do. They target a single individual. This is where the resemblance with magical thinking and ancient divinatory practices enters. Divination was based on the assumption that the future could be seen in advance and was revealed to the supplicant in strictly regulated ritualistic settings. Likewise, algorithmic predictions claim to have knowledge about the future, and by directly addressing the individual, intervene in human behaviour.

Where does the predictive power of algorithms come from? The convergence of three strands of developments have brought us to where we are now. The first strand is the unprecedented availability of and access to an enormous amount of data, aptly dubbed big data, collected not only from smartphones and credit cards, but increasingly by sensors in public spaces and private homes or implanted in AI wearables. But the reach of big data goes far beyond our behaviour as consumers or voters, or even the risk models used in financial markets. Big data now forms the backbone of many sciences, from astronomy and cosmology, which have always relied on it, to the life sciences, where computational biology and precision medicine have become unthinkable without it. Data is hardly the new raw material nor an inexhaustible resource, as its management and cultivation can be quite exacting (Leonelli and Tempini 2020).

The second strand consists of algorithms, sets of

mathematical equations and coding rules designed to carry out a specific function. Computer scientists have been working on algorithms for decades. Initially, they were designed to follow logical operations. The results were disappointing, leading to a period when funding dried up, remembered today as the 'AI winter'. It was only during the last decade, when the combination of greatly increased computing power, optimizing techniques and the availability of big data were brought together, that a new generation of algorithms started to dominate. Broadly referred to as Deep Learning, a subfield of machine learning, this second generation of algorithms is based on a simplified version of neural networks. They turned out to be astonishingly efficient, capable of creating rules themselves by interacting with massive amounts of data, with relatively little intervention from humans. The stupendous achievement of an AI system teaching itself to play chess was soon followed by an AI beating the world's best Go player, as well as other widely publicized feats. Although these are relatively narrow and rule-dominated domains, exactly how the AI arrives at these results is still poorly understood.

The third strand is the huge advance in computational power, the precondition for algorithms to detect patterns in big data from which predictions are extrapolated. Whereas the first generation of algorithms was crafted entirely by humans, the algorithms derived from machine learning are technological products, owned by the companies that design them, with many of them kept secret. Deep Learning has led to rapid industry-driven advances in AI, but not everyone in the AI community is content with the 'unreasonable effectiveness' of Deep Learning algorithms. Eventually a much better understanding will be needed of where and for which tasks each algorithm is optimally to be deployed and what its limitations are.

The digital time machine in which we find ourselves is fuelled by the predictive power of algorithms. We no longer practise divination, but we are as keenly interested in what the future holds as our ancestors were. We leave digital traces of our behaviour everywhere, traces of what we buy, eat and of who we meet. We wear fitness bands to continuously monitor the state of our health. In what Shoshana Zuboff

calls 'surveillance capitalism', we collude in delivering information to big corporations about some of the most intimate
features of our lives in exchange for more information about
ourselves (Zuboff 2018). There are uncanny moments when
we are startled by the realization that an anonymous AI
system may know us better than we do ourselves, but these
moments don't lead to any structural change. We might voice
concerns about our privacy and call for better regulation and
protection, but only end up going back to our old habits.

The advances of modern science and notably of physics
are based on the invention of new theoretical concepts
and the testing of predictions based on them. This is why
accurate predictions are still considered the hallmark of
modern science. Between the mid-1800s, when probabil-
istic prediction was introduced into physics, and today, an
interesting conceptual change in the meaning of prediction
occurred. Initially, predictions in statistical mechanics formed
the basis for a novel, stochastic view of the laws of nature.
With the discovery of quantum mechanics around 1900,
questions arose about the role of chance in the laws of nature
and how this affects predictions. Today, the interpretation of
prediction is related to the investigation of complex systems.
In this development of the meaning of scientific prediction, a
trade-off emerges between precision and the range of applicability (Hosni and Vulpiani 2017; Holovatch, Kenna and
Thurner 2017). In other words, the wider the set of problems
to which the notion of prediction is applied, the weaker
the forecasts become. While this may need to be taken into
account in science, the spill-over into society has been huge.

One field in which scientific predictions based on mathematics and simulation models have been enormously
successful is weather forecasting, which is indispensable for
transport and communication, for disaster warnings and for
preparedness around the world, including the monitoring of
global climate change. The weather prediction machine is
a great scientific and technological achievement. Its origins
go back to the beginning of the twentieth century, when,
pioneered by Vilhelm Bjerknes and Lewis Fry Richardson,
mathematical calculations were combined with the creation of
networks of observational instruments. This turned meteor-
ology into a reliable, mathematics-based science. Today,

weather predictions are based on empirical observations gathered through a global infrastructure, with a vast amount of data collected from satellites and instrumental balloons, thermometers, barometers and anemometers.

But the big shift in weather forecasting came with the advancement in computer simulation, the use of supercomputers and a purpose-built telecommunication system to bring it all together. The building-up of a hurricane in real time can now be followed on a computer screen, where the projections of its alternative pathways up to its eventual landfall allow for preparations to prevent major loss of life and property. Weather forecasting is available on specialized weather channels and on smartphones, responding to the demand for accurate forecasts because of their commercial and economic usefulness. Robots are increasingly taking over what human modelers used to do. The next challenge for predictions based on computer simulations is to move from forecasting the weather to understanding the complexities of the climate system, a challenge of enormous magnitude (Blum 2019).

Predictive analytics is most visible in daily life in commercially available products based on AIs that operate as rather simple prediction machines. Designed to make money, they promise to cut costs, while increasing efficiency and accuracy. One such prediction machine is marketed as offering 'the simple economics of AI'. Organizations and firms have to make decisions all the time and increasingly rely on predictions in order to do so. If prediction is defined as simply 'the process of filling in missing information', then every operation that takes existing data to generate new information is included – a task performed by human experts in the past. Once predictions are made by an AI, they become cheaper and more accurate, so business models change and experts are replaced. The firm's productivity is enhanced, together with its profit-maximizing strategies and task allocations. The success of the rapidly expanding platform economy, into which the prediction machines of individual firms are integrated, vindicates this logic (Agrawal, Gans and Goldfarb 2018). Not much consideration is given to external costs, nor to other downsides of the platform economy.

Another example of the application of prediction machines

comes from the creative industries. For many artists, working with digital technologies has become a regular feature of their work. They deploy AI creatively to experiment with it and to enhance their own creativity. But AI can also be used as a prediction machine to boost the marketing of the arts. In a hugely successful public relations exercise at the Rijksmuseum in Amsterdam, a group of art historians, material scientists, data scientists and engineers spent eighteen months producing what they boldly advertised as 'taking on a controversial challenge: how to teach a machine to think, act, and paint like Rembrandt'. The latest sophisticated digital technologies were employed to demonstrate that AI has advanced far enough to be able to predict what the 'next Rembrandt' would look like as if he had painted it himself.

The digital art historian Alison Langmead has deconstructed this claim. She calls it a case of computer magic, because the entire performance resembles a stage magician's trick. The algorithm was trained on selecting traits from the many portraits and self-portraits Rembrandt was famous for, and the 'next Rembrandt' emerged from this composite mixture. Generated through admirable technological sophistication, and accompanied by the organizers' claims that the AI's would match or even surpass the artist's creativity, the result was based on an extrapolation of the artist's painting style averaged across the different phases of his career. Thus, the proudly showcased AI-generated Rembrandt is not really a prediction of Rembrandt's creativity, but presents a somewhat arbitrary cross-section of different faces painted in different contexts (Langmead 2019).

If we were to set up a real 'creativity test' instead of a flawed 'next Rembrandt test', what kind of criteria would an AI system have to meet? The answer lies in the ruptures and discontinuities that mark the work of great artists and that allow us to recognize how the tensions within an artist's work evolved and were negotiated. Creativity erupts where the unpredictable enters into the work of art. Often chance plays a role in artistic endeavours as much as in science, where it is called serendipity. Randomness can also be deliberately introduced into an AI system as a way to make it learn how to be 'creative'. But as long as the machine is unable to predict a discontinuity that manifests the next step or initiates a new phase in the maturation of artistic creativity, it will not pass the test. It will remain a less inspiring example of what AI can extrapolate from Rembrandt's past work.

A rupture of a different kind occurred when modern science, beginning in the seventeenth century, started to unfold its creative potential. It had been driven by the belief that it would lead to a general betterment of the human condition, but only when technological change was introduced on a wider scale did the pace of change in the experience of time begin to accelerate. Mechanization and new means of transportation expanded mental horizons and diminished geographical distance. New and radical ideas were generated and flourished during the Enlightenment. Combined with political demands for individual liberty and religious tolerance, these ideas and the social movements that promoted them swept across Europe and radiated beyond. They led to dramatic changes in political and social life that spanned the period between 1750 and 1850. Spearheaded by the French Revolution, these developments included the dissolution of the estate system, the early impact of industrialization, and an altered consciousness of history. For the historian Reinhard Koselleck, this brought about a major conceptual change that marks this period as a *Sattelzeit* (saddle era), akin to a watershed on a mountain ridge that separates waters flowing in different directions. In this case the watershed was a conceptual change in how the future was conceived and experienced. For the first time the future came to be seen as an open horizon (Koselleck 1979).

The gap between the 'space of experience' and the 'horizon of expectation' began to widen when a fundamental difference appeared between the past and the future. The experience of the past with its slow pace and restricted life chances – limited to inheriting those of the previous generation – gave way to the idea that people's lives could be different, generating aspirations for a future that would diverge from the past. Unheard of changes in mentalities and social behaviour followed. The novel concept of the future as an open horizon made it possible to escape life conceived as a predetermined destiny and to grasp that one could shape it. It was a great discovery that turned into a great social invention.

The radicalism of this conception of an open future must be seen against the vast historical backdrop of millennial cosmologies, religious prescriptions, collective imaginaries and the lived experience of millions of people who believed their lives had to be lived as an inescapable fate. Backed by modern science and technology and the systematic exploration of the natural and social world, the idea that humans can shape society and their destinies eventually gained ground. The feeling of being at least partly in control of the future through planning, propelled by technological progress and social engineering, reached its climax during modernity. However, it turned out to harbour dangerous illusions of being in complete control, especially when the centralized state assumed such a role (Scott 1999).

Yet, we may soon find ourselves at a new crossroad. The planning frenzy of modernity had already had to yield to a humbler management of uncertainty when predictive analytics came to the apparent rescue, promising the objectivity and efficiency of employing algorithms instead of relying on fallible humans. But the more human decision-making is transferred to predictive algorithms, the more power they will exert, until they become firmly rooted in the social fabric of society. The return to a deterministic worldview becomes a possibility again. So far, many applications of predictive algorithms have been geared towards the promise of a shiny, commercialized future, of the kind people have become accustomed to crave. But they have also been quickly taken up to replace human decision-making in the delivery of public and private services, in the decisions made by courts and the police, by insurance companies and in healthcare systems.

Life in the digital time machine confronts us with a paradoxical situation. We have very efficient instruments at our disposal that allow us to see further into the future, covering the dynamics of a wide range of human activities as well as natural phenomena. Their efficiency is so convenient and economically highly profitable that it seems to eclipse the need to better understand and improve the mechanisms that underlie them. We entrust them with our most intimate data, but then worry that our privacy is being further eroded.

The paradox arises from, on the one hand, the capacity

of predictive algorithms to make happen what they predict, and, on the other, the fact that attempting to predict the future threatens to close its open horizon. Once it has circulated widely, a prediction that was intended to cope with the uncertainty of the future can quickly transform into a certainty that may turn out to be illusory. There has always been a productive tension in science between advancing the understanding of phenomena and building instruments to test theories and predictions against what happens in the real world. The power of predictive algorithms is no longer confined to science, yet the gap between their instrumental efficiency and our understanding of how they work and impact our lives persists. If anything, it has become more urgent.

If we abandon the human desire to know why and to understand what holds the world together, we risk creating a closed and deterministic world run by efficient prediction machines whose inner workings remain obscure and whose impact on us goes unquestioned. Such a determinism renounces the inherent uncertainty of the future and replaces it with the dangerous illusion of being in control. Eventually, we risk being transformed into prediction systems ourselves. Even our ability to teach others what we know and share what we experience might begin to resemble that of a prediction machine. A return to a deterministic worldview would imply that the open horizon of the future is closed again. It would mean abandoning a precious and hard-won discovery that was made only a few centuries ago. Instead, we should embrace uncertainty, allowing it to lead us into the vast space of possibilities and thereby enrich what we know.

Life in the digital time machine has many facets. It is not risk-free, even if it allows us to detect some risks earlier. It is fuelled by prediction machines, of which there are many different kinds now at work. Some are simple, but very efficient. Many have been designed to make money or to save it. Some have doubtful or sinister purposes built into them. We have to learn when to trust them, and what to entrust to them, if we want to keep the horizon of the future open.

Meanwhile, our scientific-technological understanding of the world of complex systems advances. Analysis of such systems enables us to follow processes that may lead to the emergence of new properties, and to identify the tipping points that precede a transition, leading to possible collapse. Looking ahead in this way may buy us time to be better prepared and to strengthen the resilience that resides in networks. It gives us a chance to act while there is still time to do so. It allows us to see that there are always more options ahead. Life in the digital time machine can show us that alternative futures exist and the future can be otherwise.

2
Welcome to the Mirror World

A mirror world in the making

The impact of digital technologies is manifest in an altered experience of time and a growing awareness of the spatial interconnectedness of the Anthropocene. It affects our lives, our culture and our relationship to each other. The question of what it means to be human arises repeatedly, triggered by our ever more intimate and intense interactions with our artificial digital creations and creatures. This chapter sets out to probe some of the mechanisms that shape and channel these changes. They reside in the uncanny ability to utilize the information available in and about the real world to create a virtual world populated by avatars of objects and people, by digital twins and sprawling complex systems that can help to generate useful knowledge out of data.

The mechanisms that reshape our identities and our relations to others and the environment emerge from the interactions between the virtual and the real. Any intervention in one triggers reactions and feedback in the other. Whatever we shape in one is re-shaped in the other and vice versa. Information becomes action and action turns into information. We are creating a mirror world, a world that quickly becomes our extended virtual habitat. The mirror world is not a replica; it does not consist of mere copies of an original. During industrialization, mass production led to the manufacturing of thousands of copies, which generated the cult of the original as a reaction. 'The culture of the copy', as Hillel Schwartz called it, spawned dreams and nightmares of turning one into many (Schwartz 1996). Stories about multiple personalities and of romantic *doppelgängers* proliferated. Although we have left the culture of the copy behind us, the agonizing anxiety involved in distinguishing the fake from the authentic has re-emerged as digital technologies have rendered the two almost indistinguishable. Since the digital objects in the mirror world are not copies, copies no longer have an inferior ontological status. Instead, we struggle to define the status of digital objects.

By externalizing knowledge and information in a cyber-physical environment we have greatly expanded the scope and range of intervention in both worlds. The digital tools, devices, objects and processes we have created function to connect them. They serve as go-betweens, triggering multiple feedbacks, actions and reactions on all sides. Our world has become digitally interconnected, but the vast digital infra-structure we have built, with its extended sensory networks, connects us also to the mirror world. Let us briefly look at some examples that have become familiar.

Everywhere citizens have become accustomed to the rapid spread of surveillance cameras in public spaces. They promise security. Meanwhile, smart home devices are marketed as 'helpful' in further improving people's lives. Mindful of growing concerns about privacy, Google Home cleverly extends the concept of privacy from a solitary to more inclusive notion, allowing us to accommodate not only personal but also the communal experiences that we want to share. Deciding who and how many others we want to share with becomes an optional feature built into smart home devices. Telecameras in living rooms come with a small plastic window that can be blocked, while the most recent domestic assistant, Google's Nest Hub Max, is equipped with a video camera whose additional features enable facial recognition. It comes with the assurance that this personalized facial recognition capability has nothing in common with that of public surveillance cameras as it is similar to the one already installed in iPhones (Cau 2019).

Another well-established digital infrastructure, widely used in retail shops, is the Radio Frequency Identification tag (RFID). Its promotion suggests socio-technical imaginaries of perfect efficiency and order as tagging permits four-dimensional traceability and control of the logistics that cover the movement of all the items in a shop. What remains unsaid and largely unnoticed is the fact that the identification tags are not removed from the bought items when the customers leave the shop. The algorithmic control that drives the choreography within the confines of the shop does not (yet) extend to the world outside. The algorithms remain largely invisible and unquestioned (Felt and Öchsner 2019). But once this RFID infrastructure is further expanded into the world outside, it will be another extension of our world into the mirror world.

These everyday examples serve as an illustration of how digital infrastructures are building the mirror world. In principle, every item or entity in the physical world, every event, phenomenon or object, can be given a digital shadow, counterpart or twin in virtual space. With ever greater resolution, the artificial eyes of the sensors blend easily into the urban and rural environment, be it public or private. Digital super-vision exceeds the advantages offered by digital super-speed, as cameras and sensors are equipped with software to monitor, update and intervene in what is happening anywhere in reality. One of the effects on humans is that not only do robots 'see' the world, but we begin to see it through the eyes of robots (Kelly 2019).

Thus, a huge monitoring and mapping operation is under way, from city streets to the depths of the oceans; from surveying the urban underground to following the minute operational processes in microsurgery or in delicate manufacturing procedures. Drones and satellites cover what is to be seen from above, while sensors penetrate the surface to map what lies underneath. Building digital models of the physical world in 3D brings many benefits. In a mirror world, immersive virtual replicas can be used as trouble-shooters for malfunctioning machines or in locations that are difficult to access. Repairs can first be practised in virtual reality making them easier to execute in reality. Already in the 1960s, NASA engineers were keeping a duplicate of every machine sent into space. Prototypes no longer need to undergo time-consuming and costly tests in the real world, since they can be tested in a virtual environment according to specified criteria. Meanwhile, the world's military powers attempt to outcompete each other in developing automated weapon systems. Drones can kill anywhere and at any time – perhaps the ultimate intervention by digital objects.

Based on computer simulation, digital twins can be endowed with volume, size and texture with the aim of enabling them to act like avatars. The genetic information stored in each of us makes up the digital twin of our biological self. These twins are not copies of an original. They can rightly claim to be the digital original. Biobanks around the world are rapidly filling with samples from human DNA and tissues of all kinds, with frozen oocytes and sperm and other spare parts of the body. A subset of the mirror world is created, and not only to serve as a reservoir for biological replacements; it is a laboratory for altering and reproducing life, combining and recombining what nature has made. We use the mirror world to improve the work done by biological evolution over thousands of years, and are thus actively engaged in pushing our homemade cultural evolution further and further.

So far, the scope of the mirror world is limited by the finiteness of planet Earth. However, mirrors of a very special kind are positioned in the Atacama Desert of Northern Chile, where the world's Very Large Telescope (VLT) facility is installed. Each of the four telescopes has a primary mirror 8.2 metres in diameter with an angular resolution of 0.002 arcseconds. By a parabolic arrangement of the mirror segments and other measures, an astronomical interferometer array obtains impressive results from events in faraway galaxies. Here, the mirrors still retain their primary function of reflection. They are integrated into a larger technical facility that brings the deep past into our present. Our own mirror world is but a tiny speck in the immense universe, but mirrors are also part of our continuing efforts to explore the outer space around us.

Mirrors have been used since ancient times, when humans first discovered that they could fabricate a smooth metal surface to reflect an image. Already then, they served different purposes. One of them was to gain self-knowledge. In Greek mythology the legend of Narcissus comes in two versions. In one, upon seeing his image reflected in the water, Narcissus falls in love with himself. In the other version he had a twin sister who died. Looking at his own mirror image, he longed to see her again. Today, we surround ourselves with the mirror screens, big and small, that proliferate in the digital world. Apparently, we cannot stop staring at ourselves. We are never quite sure whether we are looking at our true authentic self or at a self fabricated by social media. Maybe this adds to the excitement teenagers feel when using apps that allow them to lip-sync and dance to the same tunes. It is easy to get lost in the hall of digital mirrors where celebrities and influencers gain millions of followers while making a lot of money. It remains difficult to find out who we 'really' are.

'Having fun and making money' was also the slogan that excited the pioneering generation of young computer nerds, smitten by the new technology. They were passionate about being at the forefront of a rapidly developing computer business, driven by the belief that their generation would shape and take over the world. Caught in the amazing exponential growth of an industry that followed Moore's law, as computers reduced dramatically in size while increasing their global spread they rode the waves of success, fame and money. They had fun and the luck of being born at the right time and place. They were also full of dreams of emancipation that resonated strongly with the counterculture of the late 1960s. But the industry has since long outgrown its juvenile enthusiasm, and abandoned its ideals on the way. With it comes the painful regret 'I wish we'd built a better industry', as confessed by a well-known CEO of a US software services company (Ford 2019).

What is the response of today's younger generation to the digital world in the making? Having fun and making money retains its allure. The number of those eager to enter the start-up scene and to build the digital mirror world is still growing. Money, fun and power are the chips in the game, and gaming and e-sports have become hugely popular. They are today's mirror world for the younger generation. A world one can easily slide in and out of, meet friends and chat, play as part of a team, show one's allegiance to the Club by buying into all the paraphernalia, from which the real money is being made.

With an estimated 2.2 billion users worldwide – a quarter of the world's population – the gaming industry occupies a comfortable position, whether it is based on free to play or pay to play, multiplayer gaming or single player. Electronic sports and video games have vastly exceeded traditional live sports in popularity. They are highly competitive, with professional leagues playing for packed stadiums, from which there is much money to be made (*Economist* 2020). The first-ever *Fortnite* World Cup, for example, attracted more than 250 million users globally, among them e-sports professionals, and huge numbers of concurrent viewers on streaming platforms. With the prize money totalling 30 million US dollars, the twenty-six-year-old top winner walked away with $3 million, while millions of unknown teenagers looked up to and celebrated the *Fortnite* celebrities.

Games like *Fortnite* represent an encounter of the mirror world with the real world, where fun is shared by the players and money is made by the companies that own the games. As for mirrors, *Battle Royale* – a rather simple game in which the winner is the one who eliminates the ninety-nine others who have been dropped together on an island – holds up a mirror to contemporary society. It is still preferable to a real battle. In the classic Chinese novel from the sixteenth century, *Journey to the West*, the Monkey King looks into the magic diamond mirror and discovers he can distinguish good from evil. We still struggle to define the fluid boundaries between good and evil and to determine what in the digital world may turn out to be a blessing and what a curse. Yet we continue to use mirrors, digital ones and others, to look and reflect. We ask questions that are mainly about ourselves and expect to receive answers that tell us more about ourselves. Looking into the mirror should reveal something that transcends the image that is reflected. Perhaps it is consciousness, the evasive and unresolved puzzle of the lived experience of existence, or perhaps what used to be called the soul. We continue to search for our identity and who we are.

Developmental psychology has had a long-standing interest in the study of consciousness, which it has pursued by using mirror tests for self-recognition. Human infants recognize themselves in a mirror at some point between the age of fifteen and twenty-four months. This has been adopted as the reference point for the mirror test intended to probe whether animals are able to recognize themselves. It soon turned out that it was unclear what self-recognition in animals actually means. Does it involve a consciousness for which no scientific consensus exists? Self-awareness or a continuum of self-cognizance? Experiments with primates in mirror tests have found some sort of self-awareness and perhaps even awareness of the perspective of others, but the results remain tricky to interpret.

With other animals the influence of the environment must be taken into account. Fish, for instance, live in a very different environment from that of primates or elephants. Each species has a different perspective on the world, a unique *umwelt*, as Jakob von Uexküll, in his pioneering studies a century ago, called the subjective-temporal worlds in which animals live. Von Uexküll was a biologist who was interested in animal behaviour and anticipated much of biocybernetics. The *umwelt* is the environment that an animal species perceives in accordance with its own cognitive-perceptual apparatus, taking the relative lifespans of different species into account (Von Uexküll 1909).

It turns out that not only does the physical environment of animals differ, so too does their *social umwelt*. Dolphins, chimpanzees and orangutans are all social animals, and so are cleaner wrasses, a fish often used in mirror experiments. Do they all have the same or at least similar cognitive facilities when it comes to recognizing themselves and other members of their species? (Keenan, Gallup and Falk 2003). The mirror test has been further adapted to include cultural traits in the behaviour of social animals. Instead of showing only the face, which is a typically anthropocentric approach, larger mirrors have been used showing the whole body, and in some tests animals could see their partner. Again, important differences between species came to the fore. Gorillas tend to evade eye contact and elephants like what they smell and hear more than what they see. Their sensory repertoire is broader and certainly different from humans, who privilege the eye (Preston 2018). It seems that humans, including researchers, evidently struggle to place themselves in the position of other species and adopt their viewpoint. We remain a very anthropocentric species, in need of mirrors to open windows into our minds.

One such window opened up unexpectedly when I moderated a panel discussion at Nanyang Technological University in Singapore, shortly before the Chinese New Year in 2019. My discussants were Michelangelo Pistoletto, an internationally renowned artist, and Ben Feringa, the Nobel laureate in chemistry. It turned out that mirrors played a central part in the work of both. Pistoletto discovered mirrors early as a young artist, not in a self-absorbed, narcissistic play with his identity, but as a gateway to the social world. Instead of closing in on himself using mirrors for self-portraits, they became a means for portraying the social world. The artist, he argued, discovers the other by seeing the self: 'The identity of my fixed image tallies with the identity of any other person who, looking in the mirror, carries out the same process of establishment of an identity as I did. Each of us, looking in the mirror, can examine the whole of physical existence that lies in front of the mirror' (Pistoletto 2016).

Ben Feringa's lab in Groningen works with and on molecules. Most of the work is built around the concept of chirality, which plays a central role in the lab's research on the design and synthesis of molecular machines, for which the Nobel Prize was awarded in 2016. An object or system, in this case a molecule, is chiral if it is distinguishable from its mirror image. It cannot be superimposed on it. The term is derived from the Greek *kheir* (hand). Mirrors are by definition linked to asymmetry. A symmetrical object viewed in the mirror looks identical to itself, but only asymmetrical objects can be mirrored. Take human hands. An ambidextrous person can write with both hands using the same pen. However, using right-handed scissors with the left hand does not work well – to cut efficiently using the left hand, the mirror image of a pair of right-handed scissors is required. In the work on molecular motors in Feringa's lab, chirality is applied in a very practical sense by using the chiral centre in a molecule to enforce the directionality of the motor's rotation. If the motor rotates in a clockwise direction, its mirror image will rotate in a counter-clockwise fashion. The practical application of this knowledge extends to drug manufacture and many other fields (Lubbe 2019).

Both art and science embody the cultural evolution of the human species. They are driven by an insatiable curiosity about how the world functions and they open up new understandings. Scientific advances overturn the sensory first-person impressions we have and reveal other layers of an extended reality. Through technological invention we intervene and change the world, as with the mirror images that make molecular machines turn in the desired direction based on the understanding of how chirality functions. Likewise, artists make us see the world from a different angle. This can be seemingly as simple as seeing the similarity with others in the mirror when one looks at one's image. Artists also use and depend on technologies to make us see things differently. In each artist's work, there is a chiral centre somewhere, an asymmetric centre which ensures that the object and its mirror image are never completely identical. The mirror of the artist's reality remains asymmetrical, an invitation for the viewer to see themselves and others in unexpected and novel ways.

Our digital mirror world in the making may also have a chiral centre somewhere, ensuring this non-identity of the object and its image. The mirror world is the latest cultural milestone in the externalization of knowledge that began with language and communication in oral form and was subsequently complemented by writing. The enormous impact of the printing press enabled the rapid spread of ideas through words and images through the hierarchical layers of society and across the globe. The era of digitalization has been marked by the advent of the internet and digital media, setting in motion another dynamic swirl of diffusion. This time, however, the externalization of knowledge is no longer restricted to words and images alone, even if their communication always depends on material carriers and networks. This time, we are creating a mirror world that contains digital entities built to interact with us and to intervene in our world.

Before, the world of knowledge was contained in books and hosted in libraries – physical spaces reserved for a well-ordered and coordinated access to all available knowledge.

Libraries opened up a world that was imaginary but also real in the sense that it establishes links with the real world that are to be further explored (Connor 2019). In contrast, digital externalization allows us to build real objects that intervene directly in the real world. There is space for enormous amounts of data, information and knowledge to be stored on tiny chips, data that can be processed to find patterns or to simulate the dynamics of complex systems. The mirror world functions as the externalization of knowledge through digital technologies. It becomes also our world, just as libraries opened up knowledge of the world beyond the physical space that housed the books. We interact with the mirror world and it intervenes in what we do. It has a major impact on our identities and who we think we are, just as the first uses of metal surfaces as mirrors must have had on our ancestors. Welcome to the mirror world that reflects what we do and who we are. Welcome to our digitally extended world.

Humans and their digital others

As millions of Chinese viewers watched with awe and dismay the defeat of the world champion in their beloved game of Go by AlphaGo Zero, it must have been heart-breaking. According to Kai-Fu Lee, it also functioned as a wake-up call in boosting China's efforts to become an AI superpower. Viewers realized that there were others who were better, stronger and more intelligent. This time, they were not the Americans, Russians or Koreans, but a machine that could come up with solutions no one had discovered before in a game that had been played over centuries in their homeland. Since then, China's lead in the AI race has rapidly advanced. In contrast to the strategy pursued by Silicon Valley, focused mainly on information gathered from online behaviour, Chinese companies, as well as the government, gather data from the real world, about how and where people move, what they eat and physically purchase. China's advantage, Lee concludes, lies in its physically grounded technology ecosystem that gives its algorithms many more eyes into our daily lives (Lee 2018).

We have all become accustomed to being surrounded

by digital machines and to the idea that they may one day outsmart us. But despite their omnipresence, they remain elusive entities. They allow us to do things that had been impossible before, but many of their operations remain invisible. They fill our collective imaginaries, but we have no clear idea how they actually operate or what and who they are. We continue to interact with them in multiple ways and are aware of – perhaps fascinated or frightened by – how they affect our behaviour and our outlook on the world. As with earlier scientific and technological advances, they give us access to a new world that expands what we see and can do. But who are these new co-fellows that have increasingly invaded our living space?

Let us first look at robots. They are present in the outer environment, inside our bodies, and next to us. Roberto Cingolani, in whose lab robots are actually built, is adamant about who they are: they are another species. They clean up toxic waste, deliver drugs to inner organs and replace parts of our body. The progress achieved in mechatronics, a combination of mechanics and electronics, has enabled the construction of robots with advanced biomechanical and sensorial capacities. This has required close collaboration with other disciplines like the neurosciences, mathematics, psychology and linguistics. Thanks to highly sophisticated control algorithms and supercomputers that allow an enormous amount of data to be visualized and analysed, we have managed to create 'intelligent' robots that can move on their own, learn and perhaps even take decisions. In this, they are similar to us, but also profoundly different.

In humans, mind and body are profoundly interconnected and synergetic. Neither has a dominant role and millions of years of evolution have optimized their reciprocal adaptation, mediated through the biochemistry of life. So far, it has been impossible to transfer the inseparable connection between body and mind to the machine. The mechatronic mechanisms on which machines run differ fundamentally from those characteristic of living organisms. They necessitate energy that is orders of magnitude higher than what is consumed by living beings. In an intelligent machine with a humanoid or animaloid body an electronic programme computes complex algorithms which generate digital signals, commanding the amplifiers and switches that send the electric current into the motors that animate the robot. Compared to their biological equivalents, these processes are very rudimentary and costly. Robots follow the laws of electricity, while living beings follow those of biochemistry.

Given these fundamental differences and constraints, robots will most likely remain individually 'stupid', with limited computational capacities but still able to move around with agility and to interact physically with the world. The cognitive part of the robot, however, will have to be managed in a different way: with a kind of global repository of the intelligence of all the machines, a single 'mind' to which all robots are connected. This is where all information, including what the robots learn, will be stored and continually updated, and where each of them can upload and download their individual experiences.

The difference between them and us thus comes down to each of us having an intelligence and memory of our own, while a single intelligence will likely be shared by robots. Humans are autonomous individuals constituted by a body and mind that work in synergy, while robots will have many bodies connected to one collective mind, similar to what is called the cloud or the internet today. They will have a proper but shared intelligence to which all individual robots contribute. This is what makes them another species, one that has no equivalent in the world of biology, but which we will have to learn to live together with.

Ultimately, the difference between us and them rests on the distinction between life and non-life. Cingolani reminds us that the evolution of life depends only on six chemical elements: carbon, oxygen, hydrogen, nitrogen, calcium and phosphorus (Cingolani 2019). The remaining elements of the periodic table are used by us to invent things that nature has not created. Playing with the infinite possibilities that the combination of the other 112 elements offer has given us nanotechnologies, and will be the basis for everything else yet to be invented that is not biological. It lies at the origin of and is the real reason for our success in creating another species. It rests on the insight that life cannot simply be reproduced by us, but that we have the ability to create something that differs from it.

The realization of these limitations opened up new pathways and was crucial in the progress achieved so far. It enabled researchers to dispense with the idea of trying to copy nature, as they realized that it cannot be copied. At best, it can be imitated. This insight proved to be liberating as it brought the purpose of building robots into a much clearer focus: *they are another species.* They are humanoid technologies that should be as human-centred as possible, since they are made in order to serve us. The computations and algorithmic operations governing the movements, touching capabilities and visual systems of robots are totally reliant on a smoothly functioning energy generating and supply system. This contrasts with the amazing energy efficiency of the human mind and body. Compared to robots, our performance is inferior in certain respects, but the biological solution for our energy efficiency invented by Nature makes us far more self-sufficient.

Thinking of robots as having a collective 'mind' that governs the behaviour and decisions of individual machines also helps to explain the difference in the kind of intelligence they possess. Humans are individually diverse. It is the biochemistry of life that makes each of us unique, including in our irreproducibility, irrationality, unreliability and creativity. We also differ from robots in our approach to problem solving. Robotic algorithmic intelligence is highly precise and reproducible, specialized in finding solutions to specific classes of problems, like playing chess or Go, within a set of strict rules. But this intelligence lacks flexibility. It is poorly adaptable and has no imagination. Challenged to come up with solutions, the robots' will be of a specific type and will tend to resemble each other. Human solutions are different from each other. This is the reason why it takes time to agree amongst ourselves. We need to argue, negotiate, deliberate and occasionally fight it out.

Many issues lie ahead, waiting to be tackled. Some are technical, for instance how to enable robots to interpret not only words when communicating with humans, but also gestures and intentions. The most burning issues, however, pertain to the numerous unresolved regulatory and ethical concerns: who will manage the global intelligence, the central brain or brains, of the other species? How to ensure that this will be a kind of global commons for the benefit of humanity? How to make robots compatible with humans? The possibility of abuse is never to be excluded. In the end, it is not the robots who are to be feared, but humans using robots with mischievous or evil intentions. Aligning the values designed into machines with human values must be preceded by aligning the values of corporations with those of a digital humanism. The best antidote against the fear of robots is the knowledge that we have created them, not as an aim in itself, but to serve a beneficial purpose. We therefore need a culture of interaction with the other species (Cingolani 2019).

So far, we are only at the beginning of a focused approach to creating such a culture. Spectacular publicity stunts like the case of a robot called Sophia being granted Saudi Arabian citizenship, or less spectacular ones like the sight of robots greeting hotel guests, appear like naive versions of the imaginaries of robots originating from another age. Frankenstein's monster continues to live multiple imaginary afterlives, from the scary to the grotesque, the banal to the original. Written in the summer of the infamous year 1816 – when, following one of the most powerful volcanic eruptions ever, on the island of Sumbawa in Indonesia, large parts of Europe experienced extremely wet weather and harvest failure – Mary Shelley's story about the creation of an artificial living being resonates to this day. The myth of a humanoid monster created artificially by a human is about crossing the line between species and reveals the deep-seated anxiety about human identity. Ultimately, it may be about the wish to live forever and to defy death.

Such imaginaries of grotesque monsters are at odds with the kind of robots we may need in old age, ready to feed us when we are no longer capable of doing so, remind us to take our medication or simply keep us company when we feel lonely. This is why learning to interact with them is important. It is about communication and emotions, about needs and desires, all of which are profoundly human. Empathy and anxiety, motivations and strivings, but above all caring for others, involve human feelings that robots do not and probably never will be able to attain. They do not feel like us, as they do not feel at all. When we communicate, our deep-seated instinct is to elicit a similarly communicative response. An animated conversation involves talking to someone who is equally animated by it. To some extent robots can be taught to simulate human-like responses, while we continue to project our communicative behaviour and feelings onto them.

This discussion is highly relevant when it comes to social robots and the problems faced by ageing societies in providing care for the elderly. Criticisms of such robots raise fears of a downgrading of human–human interaction, a loss of privacy and even a betrayal of human dignity. Margaret Boden for one has taken a firm stand against imagining that future AIs might ever resemble humans in having their own needs. Needs, she argues, are intrinsic to, and their satisfaction is necessary for, autonomously existing systems such as living organisms. They can't be ascribed to artifacts. An AI needs neither sociality nor respect in order to work well. They don't care and are, literally, care-less (Boden 2018).

One country in which these worries are not shared is Japan. There, a robot has been introduced into care homes and hospitals where it has significantly helped people with mild to moderate dementia, and efforts are under way to use robotic 'carers' more extensively. In Japan, robots are preferred as carers over human immigrants. They are accepted as members of society and often regarded as family members. The reasons for this greater willingness to integrate robots have been widely debated. Many observers agree that it might be related to the Shinto cultural tradition which does not distinguish sharply between animate and inanimate worlds. On this cultural continuum robots are somewhere in-between. While not exactly 'living things', they are not considered to be discontinuous with life either (Boden 2018).

Other commentators have pointed out that generations of Japanese children have been raised on popular manga series featuring robots as helpful heroes. The most famous and popular of them is *Astro Boy*, the tale of a humanoid robot created by Dr Umataro Tenma to replace his lost son. Like other heroes, Astro Boy fights evil and injustice, aided by seven non-human superpowers. Other explanations focus on the economic case for the increased use of robots across much of Asia. They do the work of humans, helping them to deal with the 3Ds: dirty, dull and dangerous tasks. Countries that have the highest adoption of robots also have some of the lowest rates of unemployment (Thornhill 2018).

But cultural explanations retain a strong appeal. 'Renting' a relative – hiring fathers for children in single-mother households, or husbands for company events, or an actress to visit an elderly couple whose daughter is far away – may seem strange to Westerners. But it can also stand for the extension of the sense of what is 'real' that is frequently encountered in Japan, where you are what you do and are defined by your role and context. The self that exists in private (*honne*) has little relation to the one that is loose in the world (*tatemae*). Authenticity does not reside in the nature of anything, but in the emotions it awakens. Hence the popularity of theme parks and the equivalence given to a copy and its original. In Japanese society things that most Westerners would deem to be inauthentic can stir authentic responses (Iyer 2019).

Such explanations about the striking way in which robots and automation are culturally accepted and appropriated in Japan compared to most Western countries affirm the observation that our relationship with technology is never about technology alone. Interaction with robots is neither straightforward nor uni-directional. It is mediated by culture and the meaning we attribute to interacting with Others, be they familiar or strange. The digital Other does not come out of an evolutionary jungle, nor is it the lone stranger who is feared because their kinship relations and social ties are unknown. In the course of evolution the process of speciation has occurred many times, and humans learned early on to create new species by domesticating animals as well as plants, where speciation is more frequent. Artificially induced speciation in the laboratory has become routine, with the fruit fly *drosophila melanogaster* serving as model organism for creating new species.

Now we have also acquired the ability to create robots, as another species purposefully designed to help us to solve problems. Yet, unlike with domesticated animals, fruit flies or genetically altered mice, we still have to learn how to interact with them. Our clumsy efforts at this should recall other cultural changes that have occurred over time. A new kind of conviviality with many other living species has developed.

Animals are not only treated as pets and companions, but as living species sufficiently close to us to be protected on grounds of biodiversity and even accorded animal rights. Over time we have become more inclusive. This is a strong reminder that cultural norms and behaviour can and have to evolve alongside scientific-technological abilities.

However, our past record in treating members of our own species is abysmal. One of its most shameful manifestations that still exists today is racism. Well into the nineteenth century, people with different skin colour or ethnicity were not only classified as belonging to another 'race', but were viewed as a separate human species, believed to be permanently distinct. Social norms and even the law decreed that these species were forbidden to mix sexually with the 'dominant species' held to be naturally superior. Genetic evidence demonstrates that the attempts at reproductive isolation of some social groups utterly failed. Different social groups continued to mix for millennia, be it through migration, conquest, sheer domination and force, cultural assimilation or all of these together.

Attitudes towards race started to change with Charles Darwin's theory of evolution, which showed that humanity was one species and that evolution had established close inter-linkages between humans and other living organisms. But this was far from ending the debate. Colonialism continued to play a crucial part by claiming the right to dominate others in the name of an alleged 'civilizing mission'. People living under colonial rule were perceived to evolve at a different rate, eventually moving up the evolutionary ladder of civilization from a primitive stage to the highest and most civilized stage, asserted to be the reserve of white Westerners (King 2019). The struggle to prove that there is no scientific basis for 'race' and that humanity is one undivided species continues. It is an ironic twist of history that fundamental rights, dignity and respect are denied to fellow human beings on an everyday basis while the discussion proceeds about the rights to be accorded to robots and digital others.

Due to their humanoid or animaloid embodiment, robots appear to be the perfect digital other. But the overwhelming majority of digital technologies are component parts of digital systems, constituents of processes and operations that remain elusive and difficult to grasp, changing form and function depending on how they are assembled. They form a multitude of digital others. When reflecting on the similarities and differences between humans and these artificial creations the discussion soon turns towards the difference between life and non-life. As the premises for the definition of life vary greatly among scientific disciplines, it is not surprising that so far no consensus has emerged. The quest for the origin of life on earth, and thus for the definition of life, continues (Marshall 2020).

Questions about the difference between life and non-life resemble the colourful view through a kaleidoscope, with its brilliant components arranging and rearranging themselves, following a slight turn in the angle of viewing. As ever more digital others come into view, the question arises: What is *digital* life? What if all life is just computation? This would dissolve the difference and provide a more fundamental basis. Ever since the unravelling of DNA as a representation of a digital code, the similarities between organic life and computational processes have intrigued many scientific minds. And the more digital computing as a powerful new technology advances, the more analogies of processes in nature that resemble computation can be found, including self-assembly, gene regulation networks, protein–protein interactions and other phenomena. They support not only the idea that digital life exists, but also the assertion that all life is digital.

The problem is that the evidence for such far-reaching claims remains sketchy. To assert, as many scientists have done, that everything in the universe is digital and computational remains hypothetical at best. Digital hypotheses are not falsifiable by experiment. At worst, these assertions are a case of 'dataism', criticized by Yuval Harari as a new ideology or even a religion in which 'information flow' has become the supreme value (Harari 2018). There are many reasons why we do not know and perhaps never can know whether the physical world is digital. One robust reason is connected to Claude Shannon's channel capacity theorem, according to which information that is passed over noisy channels carries only a finite number of bits.

In this undecided situation Edward Ashford Lee proposes to think of machines as having a life of their own (Lee 2020).

He suggests viewing digital others as Living Digital Beings, or LDBs, a term he coined in private conversation, but later dropped in favour of using the traditional term 'machines'. They share our ecosystem and coevolve with us. They have a certain autonomy, an ability to sustain their own processes and to replicate, for now mostly with our help. Lee admits that digital technologies are made differently from biological beings, but he wants to bring them closer into the orbit of living organisms. By proposing that software-driven systems should be viewed as living beings, he invites us to ponder similarities and differences. Both should become clearer. Digital entities are not defined by DNA but by software programmes; they are not made of organic molecules, but of silicon and metal. They come in a great variety. Some are simple, with a 'genetic' code of a few thousand bits, others are extremely complex. Some are intelligent, many others not. Their 'lifespan' varies greatly, from less than a second to months and years.

Like robots, they share 'bodies' with each other. A laptop computer is host to a number of programmes and servers and can be hired to form a body for any number of them. LDBs can even switch such 'bodies'. In cloud computing, tasks are often delegated to different servers to ensure better load balancing or to manage temperatures. This can be compared to some biological life forms where sex changes occur during early development as an adaptation to changing temperatures in the environment, with little-understood mechanisms preserving the sex ratio of the species. If LDBs are entities distributed across many bodies that constantly change shape and function while having only one digital mind, then the question of who controls and owns this mind becomes crucial. A realignment becomes urgent, not only between human values and ethical principles and those designed for digital others, but also between those others and the functioning and design of our institutions.

Once we see them as living creatures, many features they share with us, their organic progenitors, come to the fore. Like us, they react to stimuli from the environment. They speak to us and interact with us in many ways, just as they have been programmed to do. The range of forms they assume is also vast, from single-cell organisms with a body consisting of a single silicon microprocessor to multi-cellular ones with innumerable components, a nervous system, thermostatic regulation and air-conditioning controlled by computers to keep their data-centre bodies at an optimal operating point.

But are digital technological artefacts *really* living beings? In what sense are they alive? We are back to the definition of life on which no scientific consensus exists. If we want to avoid getting bogged down in endless controversies, we ought to view a living thing as a process, not as a state or an object. It is not matter that lives, nor do we bring matter to life. But, as the progenitors of artificial living creatures, we are engaged in the processes that produce them and enable them to do things for us. If LDBs are indeed a possible new form of life, just as robots can be characterized as another species, what does this mean for the co-evolutionary trajectory ahead of us? Will thinking of them as a new form of life – artificial but living in the sense of sharing some, though certainly not all, characteristics of life – help us to better understand how the digital technologies that pervade our culture and society will change us?

The fear that we might lose control over these digital beings persists. Our collective imaginaries are quick to generate scenarios ranging from horror visions to more benign ones of moving towards what biologists call an obligate symbiosis, in which two species become so interdependent that neither can live without the other. New and complex life forms may emerge in such a symbiosis. But before indulging in imagining such scenarios, we should remember that new forms of life have been created by humans before. Domesticating arctic wolves into dogs started a long line of animal breeding, and in more recent times GMOs have entered the world markets. The process involves getting animals and plants to interact with us in ways that render them subservient to human needs and purposes. The same holds for the technologies created by humans. Robots should serve us. Perhaps this is the reason why they are largely built to look like us. We want our digital others to remain under human control, even if they are designed to outperform what humans can do and achieve.

And yet, the issue of who is in control over whom has never been as clear-cut as it seems, likewise the supposedly neat dichotomous separation between living and non-living matter, or between animals, plants and humans. Our relation with modern technology carries anthropomorphic features. We speak to our digital devices and treat some of them as if they were our everyday living companions. We attribute agency to them which enables them 'to bite back' (Tenner 1997). Take the practice of reading. It is an interactive cultural technology that uses language in written form on a material carrier to communicate about what exists in the real world or in the imagination. It affects the wiring of our brain circuitry and is affected by it (Dehaene 2020).

Thus the key to understanding how digital entities will affect us is interaction. It is about understanding what it means to affect and to be affected. Interaction is a two-way process, even if it is rarely symmetric. According to Lee, LDBs lack the ability to engage in interaction as a first-person participant. It is highly unlikely they have a sense of self-awareness, let alone autonomy: 'If our sense of self depends on bi-directional interaction, then our sense of self cannot be separated from our social interactions' (Lee 2020: 236).

Through our multiple interactions with LDBs they have become more familiar to us. We no longer understand them to be mere objects, but have elevated them to a new ontological status. We even consider them to have a life of their own, a digital life, however vaguely defined. We realize that we have to learn to live with them, which includes acquainting them with our values. Hence the necessity for an alignment that will oblige us to be more explicit about those values. Contradictions will come to the fore, as will the importance of the social and cultural contexts in which we interact with them. Learning to live with them will be a long and arduous process in which there is as much to learn about ourselves as there is about the difference between them and us.

Identity anxieties: loss and redefinition of the self

Our interactions with digital others undoubtedly affect us in many profound ways. The sense of self is implicated in these interactions and one of the results is an increase in identity anxieties. The anxiety about losing one's sense of individual uniqueness and not knowing who one is has deep roots in the human past. Legends from cultures around the world are full of metamorphoses in which humans transform into animals and animals into humans; there are also many stories of humans being possessed by spirits, or of humans exchanging their forms with one another. Archaeologists continue to find representations of human–animal hybrids or chimaeras, also called therianthropes, like the ones recently discovered in Sulawesi, Indonesia, which are estimated to be 44,000 years old (Aubert et al. 2019).

Exploring the limits and fluidity of one's identity and reaching out to what may lie beyond has induced humans to experiment with mind-altering drugs, often in carefully constructed ritualistic settings. Today, mind-altering drugs have not lost their appeal, regardless of whether they are legal or not. They give a psychological-chemical boost, strengthening an embattled self, enhancing performance and relieving stress (Pollan 2018). Other versions of temporal identity changes, such as those in Shakespeare's comedies or in some of Mozart's operas, are interwoven with playfulness: characters fall mistakenly in love or assume false identities, in a wild and often hilarious or tragic mixture of human folly, emotions and social barriers. Social identities can be worn like masks at festivities, to suit the occasion and the environment, while covering an elusively would-be authentic self.

We can compare this rich and diverse cultural heritage – in which the fear of losing one's sense of self is mixed with fun-loving experimentation, and where masks provide a temporary escape from rigid social categories which they subversively undermine – to our contemporary digital existence in which virtual identities are chosen as avatars in games of artificial life. Social media induce a greater eagerness to gain social recognition from others, and the more our sense of self is invaded by the digital technologies that connect us at a distance, the more the self craves to be reassured of its uniqueness. While the masks worn at carnivals, plays or other festivities were removed afterwards, we cannot as easily shed the layers that enwrap our technology-invaded self. As a result of the COVID-19 pandemic the wearing of face masks has been introduced to the West. This reinforced the message that the crisis was far from over, but was resented by those who denied the virus in the name of a bizarre notion of freedom.

'Identity' has become something precious and indispensable, but also somewhat ephemeral and vulnerable. In her recollection of life during the 1970s, when France had become a consumer society, Annie Ernaux recalls: 'Identity, which until then had meant nothing but a card in one's wallet with a photo glued onto it, became an overriding concern. No one knew exactly what it entailed. Whatever the case, it was something you needed to have, rediscover, assume, assert, express – a supreme and precious commodity' (Ernaux 2008: 145). Identity has indeed become something one needs to have, but it has also turned into a source of continuous worry. The business model of social media mandates that users must be kept engaged. They are pushed to compare themselves endlessly with others, triggering the wish to imitate what others do or wear and how they present themselves while pretending to be authentic. Identity has become a must-have item in a consumer society and a pivot in the digital age where identities are questioned, contested and continuously renegotiated.

But what is identity? The philosopher Ludwig Wittgenstein coolly dismissed the passionate debates: 'Roughly speaking: to say of two things that they are identical is nonsense, and to say of one thing that it is identical to itself is to say nothing' (Wittgenstein 1922). Indeed, it says nothing, and to assert that a person is identical with some attribute he or she possesses is evidently nonsense, as no person can be reduced to a single characteristic. However, logical arguments are not enough to prevent the resurgence of the crudest essentialism that appeals to a social order associated with outward visible markers such as skin colour or genitals. Essentialism is based on the illusion that an unchanging, authentic self exists. In a circular reasoning mode, identity essentialism claims to explain why people are the way they are and therefore can be grouped into categories based on unchanging differences and likenesses (Appiah 2018).

Philosophical arguments are rarely sufficient to end a debate. Here they leave out what drives the quest for identity, which is entirely social: our identity depends on our recognition by others. This is well captured in a story from the early Renaissance, called *La novella del Grasso legnainolo*, by the Florentine humanist Antonio di Tuccio Manetti. It begins in Florence in the year 1409. The protagonist is Manetto Amannatini, an artisan nicknamed 'the fat woodworker', who runs his own workshop. Among his many artist friends is Filippo Brunelleschi, famous for designing the dome of Florence. One day his friends decide to play a cruel joke on poor Manetto, pretending that they no longer know him and that he is another person. In the end Manetto realizes that he has been humiliated for his credulousness. He decides to leave Florence for the Kingdom of Hungary where he also becomes a famous architect.

At one level the story has a clear message about the identity of the protagonist: it is derived from who others believe he is. Our identities are social and do not exist without being recognized and confirmed by others. But *La novella del Grasso legnainolo* can also be read in a wider social and cultural context. In Florence at the time, tensions existed between the aspirations of upwardly mobile artisans and merchants and the Florentine elites. Wealth was pouring into the city and one way of displaying it was by commissioning buildings and art works. Artisans, architects, engineers, painters, sculptors and others were gaining in social status, becoming recognized by the aristocracy of Florence. Their social identity changed (Prajda 2016). The story's message is still valid today: we are what others see us to be. Our identities are shaped by the networks through which we are connected with others; and, increasingly for us, these networks are digital.

Similar cruel jokes are still played today, above all on the internet. Cyberbullying causes much distress to the victims, who often are children and adolescents, while parents and teachers feel helpless about how to respond. Sophisticated technological advances in producing 'deepfakes' that replace a person's real voice or face with someone else's are an increasing cause for concern, as is the rise of disinformation and 'gaslighting', i.e. psychological manipulation to erode someone's sense of self and sanity. As mistrust and disinformation spread in society, we may soon reach the point that Hannah Arendt warned against in her analysis of totalitarianism, foreshadowing the emergence of today's flooding of the internet by fake news. Once the world had become incomprehensible, she wrote, people 'had reached the point where they would, at the same time, believe everything and nothing, think that everything was possible and that nothing was true' (Arendt 1951).

We do not live in a totalitarian society today, even if worrying signs are mounting that liberal democracies are not as secure as was once believed. However, questions about identity have become a focal point of assertion, confusion and political contestation. While the assertive re-definition of identity served as a rallying cry for social movements in the late 1960s, with the feminist and the civil rights movements, it has now become a political weapon turned against those who still fight for the right to be treated with dignity. What was then a subversive idea – reappropriating one's identity against the attribution of an identity by a repressive majority – has become an embattled concept in an increasingly polarized and fragmented society. The rise of xenophobic nationalism and populism that has swept across Western societies, and the backlash against the perceived failures of liberal democracies, have been blamed on social movements which continue to fight for social recognition through so-called 'identity politics'.

The concept of one humanity was an invention of the European Enlightenment. It was based on the realization that humans have more in common than they do differences between them, that they are born equal and have a right to dignity. Enlightenment thinkers started the battle to abolish slavery, supported equality between men and women, and voiced concern about the situation of the Jews. Seen from today's perspective, the ideal of universalism was rooted in a Western-centric worldview, and its spread around the world paradoxically led to the exploitation and exclusion of subaltern populations, be they women, LGBT people or the colonized. But despite dismal failures in actual practice, the Enlightenment ideal also spawned emancipatory movements and progressive legislation, and paved the way for establishing a scientific worldview which was to play a central role in the processes of modernization.

Biological advances over the last 150 years have repeatedly changed who we think we are. But at times science, or rather scientism – the belief that science is the best and only means by which society should determine its norms and values – has colluded with the interests of politically or socially powerful groups. Such was the role played by psychology in transforming the IQ test from an educational tool into a means of social control, which paved the way for eugenics. The support of science for eugenics remains one of the dark chapters in the history of science. Attempts at 'improving' human heredity ultimately became a murderous weapon deployed to control and eliminate those who were unwanted. The spurious 'science of race' culminated in horrendous crimes against millions of people who were declared 'worthless' and stripped of their humanity (Comfort 2019).

It is important to acknowledge that science has not always lived up to its proclaimed scientific values and its own ethos. The history of science is littered with cases that reveal an uneasy relationship between scientific theories and the dominant thought of the time. It often starts with the careless use of language and of analogies that ease the transmission of ideas while stripping them of the context in which they arise. When, for instance, the sequencing company 23andMe asserts on its DNA testing kit that 'the sequence is the self', a relapse into genetic essentialism is near. At other times, cautiously articulated and differentiated scientific findings are wilfully ignored when the scientific 'message' is perceived as supporting a political agenda or ideology (Ball 2019).

Feminist scholars have repeatedly analysed claims and findings made by scientists that resulted from sloppy and outright bad science. Yet such pseudo-scientific views have been eagerly seized upon by parts of society in order to legitimize claims for innate superiority or to assert the 'natural' inequality between men and women. Many such claims have been shown to be false and untenable, yet they persist as powerful fictions that shape social behaviour (Fine 2017).

In 1863 Thomas Henry Huxley raised what he called 'the question of questions', namely 'Man's place in nature and his relations to the Universe of things'. The changes that have taken place over the past decades have huge implications for redefining our place in nature and our relationship to ourselves. Nature has ceased to be something that only exists outside of us, but is now understood to be inside of us as well, making us an integral part of it. This has far-reaching consequences also for the moral authority that was previously bestowed upon Nature. The dichotomy between *is* and *ought*, between a natural order accepted as given and serving as an unquestionable reference point for a social order based on arbitrary power relations, no longer holds. It remains to be seen which new authority will take its place and whether humanity has any choice other than to emancipate itself definitively from a fictitious external moral authority that sets the norms for what is 'natural'.

Science continues to generate fresh insights that overturn many previously held beliefs. It takes the lead in the redefinition of the self and it remains to be seen how quickly society will follow. We are not really us without non-human cells like the bacteria that live in our microbiome. Epigenetics has complicated previous notions of individuality as it dissolves the boundaries of the self by including past and possibly future generations. Biotechnology and information technology contribute to the sense of a more dispersed and distributed self, with human–machine interfaces and neurotechnical devices that extend the self into the domain of the artificial. The brain of amputees, who for various reasons reject wearing a prosthesis, can learn to accept an artificial limb as being part of the body by forming a new 'prosthesis category'. It then recognizes the artificial limb as a tool, which it is (Maimon-Mor et al. 2017).

These and other scientific advances lead to a redefinition of the self as relational. Immunology no longer defines the self in absolute terms. Tissue-graft rejection, allergies and autoimmune reactions that used to be described as a 'war', have been recast in terms of a crisis of self-recognition, hence as an identity crisis. Cell and molecular studies further relax the boundaries of the self. Reproductive technology, genetic engineering and synthetic biology have shown that human nature is much more malleable than previously thought. The increased number of sequenced animal and plant genomes show their many common evolutionary trajectories and shared molecular mechanisms (Comfort 2019).

Thus, the multifaceted concept of identity remains very much in flux. Science increasingly makes visible what was invisible before, even if visualization is still rudimentary. We can follow the movement of a single living cell in the body or detect pathways that connect the microbiome with the brain. These glances let us see ourselves in a new light. One fascinating research area, with implications for the future definition of the self, involves growing cerebral organoids in more than 100 labs worldwide. Neural tissues generated from human stem cells are allowed to self-organize and develop into a 3D conformation, which is the way the brain develops naturally (Lancaster et al. 2013). The organoids currently stop growing after a few months due to lack of blood supply to bring nutrients to the inner cells, but their growth period is likely to be extended up to one year. By following defects in the neural development process, the hope is to eventually be able to treat a range of neurological disorders (Ball 2019). Researchers are not growing brains in the lab, but miniaturized and simplified versions from stem cells in vitro. They are acutely aware of potential ethical issues, but so far no ethical downsides have surfaced (Cookson 2019). The potential of this kind of research for understanding the human brain is huge, as it is for understanding what we mean by 'the self'.

During the COVID-19 pandemic current definitions of identities and the self were challenged from yet another side. A tiny unknown virus wreaked havoc on a technologically advanced global civilization, sending entire countries into lockdowns, grounding airlines and revealing the incompetence, if not irresponsibility, of many governments. It also brought already existing inequalities and fissures to the fore, including those of socially disadvantaged groups who suffered further from the digital divide. It laid bare how deeply the self is embedded in a fragile social fabric, and how dependent our mental health and physical well-being are on social contacts and the networks that sustain them. Thus, the vigorous push towards digitalization exerted by the pandemic is also challenging our definition of the self (Nowotny 2021).

We are in the painful process of learning to acknowledge the interrelatedness of living beings, and that viruses of all kinds will remain with us. A digital self is in the making, shaped by digital technologies. Their impact on the self proceeds through various paths. Network connections already make it possible to detect individuals who are not on social media through the interconnections among friends who chat about their media-abstaining friend who can easily be identified (Garcia 2017). The stupendous advances in facial recognition by training algorithms with millions of images and photographs, as well as developments in voice recognition and the construction of artificially human voices, are part of the shaping of the self. The self is visually reproduced and reproduceable, the images in constant exchange with who we think we are and who we would like to be. A picture of one's face can now be projected forward to show the changes that will occur as we age, offering a preview of our future self. The management of one's self needs careful consideration to maintain a healthy balance.

Connecting a face with a person's identity is of course nothing new. State authorities have always been keen to identify, register and control their subjects. Later, citizens became passport and ID card holders. But now, the digitally recognized face becomes the entry point to everything known about the past, geared to predict future behaviour. Health records, including genomic data, can be combined with all other data stored somewhere in the electronic cloud. This digital twin becomes part of our new digital self. Once we lose trust in the promise of anonymity or the right to be forgotten, nothing romantic remains in the idea of a digital *doppelgänger*. Instead, we may find ourselves in the dystopia of Zuboff's surveillance capitalism. We are desperately in need of legal safeguards and digital rights that can be enforced. The social fundaments consist above all in institutions that are robust and have been designed to withstand pressure. They must also be sufficiently flexible to frame and allow space for the redefinition of the self that succeeds in integrating its biological, digital and social dimensions.

In our digital age, we need to respond to identity anxiety other than through ethno-nationalist or other forms of 'identity politics'. One space offering such an alternative model is provided by sci-fi. Just as mirrors can be mind-opening windows, sci-fi offers scenarios about the evolution of the future relationship between humans and digital others. It always revolves around the same question of what it means to be human, yet the answers vary. Let us look at some of them.

In a string of interviews, Alex Garland, the writer and director of the sci-fi film *Ex-Machina*, recalls that as a youngster he was driven to know whether computers have a mind of their own or whether machines can become sentient. His psycho-thriller deals with the question of whether a robot can convince a human that it has consciousness and whether the authenticity of attraction is an indicator. The male protagonist is a young nerd who is nudged by his boss to meet Ava. He feels attracted to her, even though, given her outer appearance, he is fully aware that she is a robot. Later it turns out that the boss, who controls the strategic AI that is Ava, is a villain. Ava passes the test of appearing human despite the knowledge that she is not, but the end of the story allows for multiple interpretations. For Ava – a combination of a genderless artificial intelligence and an android in female form – is capable of making a decision herself. In a dramatic dénouement she betrays both men, the nerd who fights for his life and his evil boss, and choses her own freedom.

Cinematic licence always permits logical inconsistencies. The ending of *Ex-Machina* gave rise to endless discussions about whether the film was three minutes too long, prodding Garland to explain why he decided to end it as he did (Reyes 2015). It all depends, he said, on who you identify with: Ava or the male hero? Both are proxies for humans and for the decisions they take. Both can perish. Garland decided to side with Ava and gave her the resourcefulness she needed to escape. Here we encounter the mirror phenomenon again: when looking at the robot, we end up looking at a mirror image of ourselves. The filmmaker's decision to let Ava escape becomes a strong statement for the indefatigable human desire for freedom.

What defines humanness is also the existential question in 'Düssel...', a short story by Ian McEwan, the author of *Machines Like Me*. The tale is framed as addressing a future younger generation who have no sense of history. As they take the extraordinary technological world into which they were born for granted, the narrator wants them to know what it was like when the first android became pregnant and the first carbon-silicon baby was born. Questions like 'Could a machine be conscious?' or 'Are humans merely biological machines?' had been hotly debated, and decades of international wrangling between neuroscientists, philosophers, bishops, politicians and the general public followed. Finally, it was decided that artificial people were to be granted full protection under a number of human rights conventions, as were their mixed-source offspring. All were given property rights and it was made illegal to buy or own a manufactured person. After all those turbulent years of anguished reflection and passion, the law had decided what it meant to be human. Humans and androids were to be treated as equals.

Leading scientists agreed that these new co-fellows could feel pain, joy and remorse. But how to prove it? The great philosophical questions eventually faded without being resolved. The new friends seemed just like humans, only more likeable. They blended well into society. Their nature was to be deeply caring and many became doctors or nurses. They also made up two-thirds of the top athletes. They showed themselves to be brilliant musicians and composers. It became socially accepted that these members of the artificially crafted con-species deserved full dignity and respect for their privacy. This meant that it became socially unacceptable and politically incorrect to ask whether they were *real*. The story begins with an exquisite erotic scene between the narrator and a beautiful woman with whom he is in love, but whose origins he does not know. Daringly, he decides to break the taboo and ask *the question*, 'Are you real?', knowing that if she is an android then she will have been programmed always to be truthful. In a hilarious scene she replies: 'I was formed in Düsseldorf in Greater France.' A love story unfolds. The beloved is a person who in many ways is superior to her human lover, something he will need to bear in mind for their future life together (McEwan 2018). Androids embody, in this saga at least, the better angels of human nature.

In the fantasy world of transhumanism, of daring escapes to remote galaxies, and of horror scenarios where an AI takes over the world and annihilates the human species, stories about beneficial artificial companions are rare. But they can be calming for the tormented mind. Humanity seems to be moving on the ridge of a mountain range where fog clouds its vision. It is known that on one side of the ridge is an abyss, full of dread, if not death; a valley of no return. On the other side is the scent of spring meadows, a whiff of paradise. With every step forward humanity could stumble over either side of the ridge, but every step forward is shrouded in uncertainty.

Taken together, all the stories we tell each other about the uncertain future ahead are part of the great storytelling treasure of humanity. Innumerable legends and myths from the most remote corners of the earth originated from people who never learned to write or read and were long thought to be primitive. Only recently have we begun to appreciate the complexity of their wonderful ways of interweaving the cosmos with the hardships of their daily existence. The treasure is also packed with stories about the cruelty humans are capable of afflicting upon each other; but, occasionally, the good wins out in the end.

Modernity has added its own versions, beginning with Mary Shelley's *Frankenstein*, which lives on with astonishing plasticity and adaptability. One of the latest is Jeanette Winterson's *Frankisstein*, an updated twenty-first-century version of what life means in the age of AI and artificially engineered beings (Winterson 2019). The quest for meaning that resonates throughout human history has remained astonishingly similar: What is *real* life these days? Even if we turn to human relationships and love as the way forward, it is unlikely that we will find the kind of answers or 'solutions' that will satisfy a technology-driven society. But we are encouraged to continue dreaming of them.

Ursula K. Le Guin, a master in imagining other worlds, put it as follows:

> Hard times are coming when we'll be wanting the voices of writers who can see alternatives to how we live now, can see through our fear-stricken society and its obsessive technologies to other ways of being, and even imagine real grounds for hope. We'll need writers who can remember freedom – poets, visionaries – realists of a larger reality ... Imagination, working at full strength, can shake us out of our fatal, adoring self-absorption and make us look up and see – with terror or with relief – that the world does not in fact belong to us at all. (Le Guin 2018: 383)

Maybe this is the digital mirror we need, a mirror that does not hide anything and that presents things in their complex interconnectedness. Maybe this is what digital humanism is all about. The world does not belong to us, but we continue to shape it and be shaped by it.

3

The Quest for Public Happiness and the Narrative of Progress

The contagiousness of narratives

Humans have always wanted to know what the future holds, but in addition they yearn for a future that is better. This desire still animates the migratory flows of millions of people around the globe who seek to escape misery and to improve their economic existence. Politicians promise jobs and more jobs while businesses advertise their products and services as making life easier and better. The market for happiness and self-fulfilment flourishes by promoting longer and healthier lives and human enhancements that will lead to more enhancement. In Greek antiquity the ideal of leading a good life, *eudaimonia*, was held in high esteem, but it also advised moderation. If the good life for an individual is to be extended to become the good life for many, their lives must be improved as well.

Beginning with the European Enlightenment the quest for public happiness set in. Food, the basic precondition of existence, was seen as the means to produce well-nourished workers which meant happiness for the nation. For the first time, from the mid-eighteenth century onward, continued economic growth took off in some parts of Europe. This was partly based on technological inventions, but not only. The economic historian Joel Mokyr reminds us that it was also novel configurations of ideas and changes in values, based on the belief that human betterment was possible, that under-pinned the rise of what he calls an 'enlightened economy'. These ideas and beliefs circulated in expanding social networks that found their hopes for betterment confirmed through the advances of science and especially the tangible results delivered through new technological applications. A 'culture of growth' had emerged that became fixed in novel institutional arrangements, foremost the market (Mokyr 2010; 2016).

In this chapter we will follow the ideas and beliefs that underpinned the rise of economic wealth and well-being, relying on modern science and technological innovation as the motors of economic growth. They recur in a narrative form, communicated through informal networks with a simple but compact message: a better life is possible. This message has been co-opted and reinforced all along by the dynamics of capitalism in its multiform and adaptive variants. A culture of economic growth would most likely not have occurred without such a narrative. The period of modernity and modernization was underpinned by the Narrative of Progress. Modernization spilled over into the most remote hamlets on earth, providing access to water and electricity. It came with some of the downsides of a modern lifestyle, like industrially pre-processed junk food, leading to obesity and other modern diseases. While the narrative of progress meant progress for all, it arrived with a considerable delay for some, while others were already moving ahead faster.

Narratives are a form of storytelling. They are about real or fictitious events, persons or processes and what happens to them. They are a potent tool of communication, enabling cooperation through what unites us or turning us against each other. Digital technologies have enabled them to circulate faster and wider, but also to make it easier to fake them. In his bestselling book *Sapiens*, the historian Yuval Harari argued that we are a species of storytelling animals – that we think in stories rather than in numbers or graphs (Harari 2014). We continue to tell each other stories about how we see the world, how we try to understand it and explain it. The origin and end of the universe can be told as a story. Stories about an unending succession of conflicts and their resolutions, often expressed in wondrous myths and legends, have been the main vehicle of transmitting tradition and culture through the ages. They are replete with heroes and villains that send clear signals to listeners about who is to be admired and emulated and who is to be condemned, shunned and banished.

Harari ends his book *21 Lessons for the 21st Century* with the passionate assertion that 'life is not a story' as *Homo sapiens* now faces a far bigger challenge. As he mentioned in one of the many interviews he gave: 'We created myths to unite our species. We tamed nature to give us power. We are now redesigning life to fulfil our wildest dreams. But do we know ourselves anymore? Or will our inventions make us irrelevant?' The inventions he refers to are digital algorithms.

Harari fears that digitalization has become a system that watches and digitalizes everything it can get hold of, and that in doing so it will usurp our thinking. His urgent plea is as desperate as it seems honest: if only we, as a species, could work sufficiently hard to understand our own minds before the algorithms make up our minds for us. If we can't, then we risk being at the mercy of sinister control commands operating in the background. This is the voice of someone gripped by the return of an archaic fear, the fear of an almighty deity that has complete knowledge of us, including our innermost desires. The passionate plea not to cede control to a powerful Other, in this case an algorithm, taps into the familiar age-old narrative of the human aspiration for freedom. Life has to be shielded against the algorithms taking over. If we let them tell stories, Harari warns, we are doomed.

Narratives cover the whole gamut of human concerns. There is the quest for *having*, our material well-being; the wish for *being*, for esteem and social standing in the eyes of others; and the desire for *becoming*, to fulfil one's potential and for human enhancement. If we cannot escape narratives, what indeed will happen when algorithms begin to tell or re-tell our stories for us? What if they follow and eagerly share narratives that have been designed to influence what we do and how to think? Worse, what if the narratives are deepfakes, fabricated entirely as a synthetic video, audio or writing that is ubiquitous and undetectable? AI can already produce shockingly human-sounding sentences, as well as transfer human voices and images into different, faked, contexts. The fact that this happens already is hardly reassuring. Narratives told by algorithms present an even greater challenge than the stories told by fellow humans: how trustworthy are they?

Trust is deeply implicated in narratives of fear. They have a long history in relation to how power is gained, how it is wielded and maintained, in the extreme through terror. When apparent certitudes break down and the public mood swings in the direction of dark times ahead, dormant narratives of fear begin to circulate. W. H. Auden's long poem *The Age of Anxiety*, written during the Second World War and published in 1947, has not been out of print since. The financial crisis of 2007–8 renewed latent anxieties, as did the migration crisis of 2015 in Europe. They turned voters against the political establishment and the elites, bringing ruthless demagogues to power by promising to make their countries great and safe again. We have apparently entered the latest phase of narratives of fear, where conspiracy theories and fake news items are circulated in order to incite people to violence.

The role of fear in political life has a long tradition. From Sallust to Machiavelli, from Hobbes and Locke to Carl Schmitt and Hans Morgenthau, narratives of fear were seen as means to impel people to act. This was expressed in President Roosevelt's famous address in 1933: citizens of the United States have nothing to fear but fear itself. Now, narratives of fear are intended to lure people into outrage and protest. They are about the loss of status and respect, of jobs and living standards, as a consequence of being overrun by immigrants. The standard cultural repertoire incites the anger, hatred and envy that are the emotional toxins threatening the future of liberal democracies (Jacobson 2019).

The work of Michele Lamont offers another example of narratives that deal with social and moral worth. They form the cultural glue that may cement social hierarchies and maintain the boundaries determining who is included or excluded from a social group. Meaning-making is central to the social processes through which unequal relationships are created and reproduced. In her empirical studies, Lamont shows how narratives interact and create moral boundaries that reinforce each other. People share concepts of worth, including self-worth, that influence and sustain the social hierarchies they help to reproduce. Thus, narratives can exercise a huge influence not only on human behaviour, but also on the classifications that uphold, monitor and sanction the symbolic boundaries between social groups, directed especially towards the less valued others. Existing residential segregation, intermarriage patterns and networks that affect access to resources, such as good schools, social capital and decent jobs, are not only symbolic, but real. Yet they are continually upheld and justified by narratives about what people deserve (Lamont 2000; Lamont et al. 2016).

The COVID-19 pandemic, like other major crises, proved to be fertile ground for narratives pretending to have discovered the truth behind what otherwise remains inexplicable. But literary fiction also enjoyed unexpected popularity. During the lockdowns, books about previous plagues were rediscovered, with Albert Camus's *The Plague* leading the bestselling list and even Giovanni Boccaccio's *The Decameron* witnessing a revival. This fourteenth-century work concerns a group of young Florentine aristocrats who, having escaped from plague-stricken Florence to an idyllic retreat in the countryside, tell stories to one another to pass the time – similar to the relief and relaxation offered by Netflix during the recent lockdowns.

But the relationship between epidemics and narratives is deeper. The metaphor of contagion, borrowed from medicine, goes back to antiquity. Likewise, epidemics are often used as a metaphor to describe the dynamics that underlie narratives and how they spread and diffuse. Ideas and narratives are transmitted and taken up selectively in a population, following specific networks. Dan Sperber, a social and cognitive scientist, pioneered the analysis of the contagiousness of cultural ideas. He charts an 'epidemiology of representations' to explain why and how some ideas are contagious and why and how some become durable and can even be transformed into institutions (Sperber 1996).

Even economics is no longer immune to narratives. Robert S. Shiller, recipient of the Nobel Memorial Prize in Economic Sciences, surprised his fellow economists at the annual meeting of the American Economic Association in 2016 by devoting his presidential address to his long-standing interest in narrative economics. Economists, he urged, should abandon their belief in the rationality of economic behaviour. They should instead explore the half-baked economic folk theories held by ordinary people and the stories they tell each other pertaining to economic events. These may be upturns or downturns, financial crashes or recoveries, which provoke theories of how everything hangs together and is related to people's everyday lives and economic worries.

Shiller's advocacy of narrative economics claims that once they are quantified, narratives are more powerful than statistics. Empirical case studies based on the quantification of narratives analyse stories told about stock market booms and busts, about the rise and fall of Bitcoin, and about various issues like the desirability of the gold standard. Crucial for this analysis is again the analogy between narratives and the epidemics of disease: both are contagious. When narratives have infected a sufficient number of people they drive economic behaviour. They become fundamental to people's reasoning in a way that affects the economy (Shiller 2019).

These examples of narratives drawn from different domains illustrate that it is difficult to escape them. Little is known about how much they actually affect outcomes, but they can mobilize, unite or turn people against each other. They are generators of meaning, offering explanations for complex situations and processes that would otherwise remain difficult to understand. They are embedded in social networks which they can create, uphold or destroy. Trustworthiness plays an important role. The yardstick for a successful narrative is not how close it comes to being true, nor what it delivers, but whether it succeeds in 'infecting' a sufficiently large number of people such that they believe in it and align their actions with it accordingly.

We will now turn to the one meta-narrative that overshadows all others, the narrative of progress, and to its predecessor in the Enlightenment's quest for public happiness. It has a fascinating conceptual history and can boast impressive practical effects before a change in circumstances initiated its current downswing. It accompanied the belief in modern science from the beginning, and was reinforced by the advances in societal improvements that actually reached people, based largely on technological innovation and continued economic growth. Its critics notwithstanding, the narrative of progress has been amazingly successful and persisted over a long period, spreading around the globe and permeating multiple modernities. Even if it struggles to transform itself at present, its influence is not yet over.

Breadfruit, potatoes and the idea of progress

The narrative of progress is deeply rooted in the Western imaginary of the Enlightenment's universalism, which blinded it to the realities of those living outside the European metropolitan territories. When, from the mid-nineteenth century onward, technology became science-based, it could claim that its achievements were intended for the benefit of humanity. The discussion soon focused on the integration of technology into the project of modernization. Progress became equated with technology on the road towards a brighter, modern future. The process of colonialization assured that this powerful narrative circulated across the globe, as the colonizers worked with local elites to use the colonies as test beds and laboratories for experimenting with technology and the social progress it would bring.

Despite its triumphant advance, the narrative of progress always had its critics and was repeatedly questioned, especially after the major catastrophes that occurred during the twentieth century. The stunning technological progress achieved in the nineteenth century did not lead to any moral distinction between what could have been achieved and what actually occurred. The sufferings that accompanied progress, and its many negative side-effects, were excluded from consideration. Technology was seen as a neutral, irresistible force and as the harbinger of a social progress that was bound to follow. The construction and handling of this force – which had tamed nature and built new communication, energy, transport, food and other systems to serve society and the nation state – were best served if they stayed in the hands of technocratic experts.

This may explain why the concept of technology as a neutral driving force of progress underpinned all the major ideologies of the twentieth century, from Marxism to fascism and liberalism, and why technocracy could attach itself to all three, leading to a shared legacy which deeply shaped the development of nation states and their agendas for social progress. It also became embedded in international organizations, many of which were founded during the last decades of the nineteenth century. It led to the idea that wars could be avoided through rational decision-making. The big shock came with the First World War, which dealt the final blow to the idea of moral progress advancing together with technological progress (Nowotny and Schot 2018).

This original hope of the Enlightenment soon had to be abandoned. Progress in the plural – *les progrès* – had no chance, as the gap between the efficiency of technological solutions and the tedious processes of getting people to cooperate, let alone improving the moral flaws inherent in human nature, was only too obvious. Technological fixes, however, have lost nothing of their attractiveness. Despite the doubts and its numerous articulate detractors, the narrative of progress has demonstrated an astonishing vitality and persistence. Given the practical benefits of tangible technological advances that have slowly percolated through all layers of society, it is difficult to maintain that progress does not occur. For some, it even advances too fast, leading to changes they have difficulty coping with.

The ingenuity of the narrative of progress consists in its tacit linking of technological progress with social progress by insinuating that the latter will inevitably follow. Technology was always seen to be the driver that preceded social progress. It offered practical solutions that all could see, thus reaffirming its lead. From the hygienic measures introduced in the nineteenth century to ensure a clean water supply, to vaccination, penicillin and a host of antibiotics today (growing antibiotic resistance notwithstanding); from the lightbulb to the washing machine that dramatically improved the lives of millions of women from one day to the next; from the horse carriage to mass mobility by air transportation – the modern world is densely stitched together by technological inventions that are intertwined with social improvements. Together they have left their imprint on our lives and made modernity a synonym of progress (Johnson 2014; Harford 2017).

The aspirations of millions of people kept the faith in modernization going and renewed the credibility of the narrative of progress. Its key message that the future will always be better than the present retained its attractiveness for a large part of the world's population, adapted to local circumstances. But the road to attaining this goal is full of potholes and it is no longer clear where it actually leads to. We have entered a period in world history when the destructive consequences of human intervention in the natural environment are making us question once again the narrative of progress. Technological progress can no longer be neatly decoupled from social progress nor from environmental sustainability. Despite the undeniable fact of overall progress in raising the quality of life of millions of inhabitants on the planet, the belief that it will continue like this is stalling.

Strong and convincing as it sounded in the past, the message conveyed by the narrative of progress is beginning to feel somewhat out of touch with actual developments such as extreme weather volatility or the Fridays for Future movement in which youngsters admonish adults to listen to what science has to say. Questions such as 'which progress?' or 'progress for whom?' are being raised, and not only by those who fear they might be left behind. Just as owning a car no longer seems to be the dream-fulfilment it once was, having access to the internet and a smartphone is no longer seen as an insurance against the risks of climate change. A more inclusive, holistic and radical approach is warranted as we increasingly face multiple and interconnected complex systems.

Has the narrative of progress failed in fulfilling the promises it once carried? If half of working-class men in the US today earn less than their fathers did at the same age, what does progress mean to them? If half of them believe their children will be worse off than they are, what does the future hold for them? Whenever lived experience lags too far behind expectations, doubts enter and the credibility of the narrative wanes. Eventually, disappointment sets in and the narrative starts to lose the grip it previously had. Trust erodes fast when the imaginary contract between the individual and society appears broken and the promised goods fail to be delivered. Some political and business leaders are ready to acknowledge as much and ask whether a new narrative is needed. Others rush to the defence of the narrative and seek to reaffirm it by insisting on its empirical validity. On the basis of historical data from the past two hundred years they extrapolate that all that is needed is to reinvigorate the belief in reason, science and technology.

This is what Steven Pinker's *Enlightenment Now* sets out to do, plotting humanity's progress onto seventy-two graphs and charts that support the story he wants to tell (Pinker 2018). On the vertical axis, the curves indicate in which of the many dimensions of progress life has improved, while the horizontal axis shows developments over time, when and how fast improvements were reached. The figures are based largely on statistical data compiled by governmental agencies and international organizations. Some critics have questioned what the figures actually show and whether violence, for instance, has really abated as much as Pinker claims. Historians have accused him of leaving out too many countertrends and counterexamples. This misses the point, however, as the graphs show general trends and the evidence is there: life in the last century and a half has improved for most people in the majority of dimensions accounted for.

Pinker's arguments rest on more shaky ground when explaining why progress occurred. The figures provide solid evidence for overall betterment, but they cannot show that this was mainly due to the ideas exchanged between a small group of interconnected thinkers during the European Enlightenment. Pinker's eagerness to defend the Enlightenment against its foes past and present induces him to be overconfident. 'The Enlightenment has *worked* – perhaps the greatest story seldom told', he declares triumphantly. The engine for human progress is a set of values that enabled it: science, reason, humanism. Today, these values need to be defended against the numerous enemies of progress, including religion, nationalism, populism, theistic morality, mysticism, tribalism, romanticism, contemporary intellectuals and all pessimistic 'progressophobes'.

Such sweeping and defensive statements, however

eloquently and passionately expressed, cannot overcome the shortfalls of the analysis. The appeal to the values and ideas of the Enlightenment is understandable at a moment when they have again become embattled, but simple causal infer- ences in history are always doomed to fail. Lauding a belief in the Enlightenment or any other set of values must also fail when there is no careful consideration of the wider economic, social and cultural conditions that converged to bring about the undisputable improvements benefiting humanity.

Let us therefore briefly turn to one of the precursors of the narrative of progress: the quest for public happiness as it arose in the eighteenth century. It offers a glimpse into what the imagined future looked like at the time, and how one of the most basic preconditions of human existence – food – was seen as the vehicle for producing happiness for the nation. Tourists strolling in pre-COVID-19 times in one of the megacities in South-East Asia may have noticed a lowly item on the menu: breadfruit. It is offered as roasted or fried; with rice or curry; spicy, sweet or savoury. While enjoying their meal, they will hardly be aware that this starch-rich fruit, first encountered by Europeans in the South Pacific in the eighteenth century, was once considered an almost limitless source of nutrition. The botanist Joseph Banks, later President of the Royal Society for forty-one years, is believed to have brought some 30,000 plants back from his voyages. He sent botanists to collect samples for Kew Gardens in London, and raved that Tahitians 'may almost be said to be exempt from the curse of their forefathers; scarcely can it be said that they earn their bread with the sweat of their brow'(Banks 1962: 341).

Such were the marvellous benefits ascribed to breadfruit that it was soon transformed in the colonial imagination from being merely a foodstuff into 'a symbol of a simple and idyllic life free from worries about work or property'. Breadfruit offered the prospect of a food source that required no labour. It thus became irresistibly attractive to plantation owners in the Caribbean, whose slaves expended extra labour on growing their own food. Breadfruit, the owners argued, could feed the slaves who would then have more time to work on the sugar cane plantations. Thus breadfruit came to represent the potent convergence of the capitalist logic of the plantation system and an enlightened fantasy of a life without labour.

This is part of the gripping story, eloquently told by Rebecca Earle, about food, colonialism and their interconnection with what Enlightenment thinkers called 'a quantum of happiness' (Earle 2017). The belief in quantification was accompanied by attempts to measure whatever could be measured. Then, as now, quantification was seen as highly desirable. Whatever could be quantified was perceived as being objective and providing welcome legitimacy to the state and its rulers. Following the same logic, various indexes of happiness were quantified and made measurable. The Scottish agronomist John Sinclair devised the 'quantum of happiness' to measure the happiness of the state by adding up the happiness of its individual inhabitants. But there were clear limits: the geographies of public happiness remained well within the realm of the cold logic of colonialism. Colonies and their inhabitants were not counted as eligible and thus remained excluded from the *félicité publique*. Enlightenment, despite its claims to universality, met its limits in the practical realities of exercising the economics of power.

The promotion of happiness was *the* great storyline throughout the Enlightenment. It was everyone's duty, the goal of the century and virtually a truism. It fell upon individuals, but even more on governments to make people happy. As both the promoters of new foods and Enlightenment philosophers observed, the most straightforward way to achieve this goal was by giving people nutritious food. The irony, of course, consisted in the idea that breadfruit represented a paradisiacal world and therefore should serve as food for the enslaved population toiling in the Caribbean sugar plantations. When planters lost interest in its dietary potential, and the intended beneficiaries were not as keen to adopt its benefits as they should have been according to the enlightened imagination of their colonial masters, breadfruit eventually ceased to play the central role it once occupied.

As Earle vividly relates, soon breadfruit trees were transformed into an adornment in British and French Botanical Gardens in the West Indies. Its voyage from being an exotic food item to becoming the symbol of beneficence, an emblem of European Enlightenment that encapsulated

Nature's bounty, came to an end. Its place was soon taken by potatoes, especially amongst the population of Northern Europe. No more seedlings for breadfruit trees needed to be transported on ships like the *Bounty*, as potatoes could be grown quickly and efficiently in European climates, serving the same purpose of nourishing the population.

Since a nation's wealth depended on a large and healthy population, and nutritious food would turn them into more energetic workers, hundreds of investigations were undertaken across eighteenth-century Europe to assess the nutritive qualities of plants transferred from the colonies. In one particularly striking example, in 1799 a group of wealthy women in Madrid began a decade-long experiment aimed at discovering the best substitute for breast milk. The ladies were in charge of the city's foundling hospital; when they were unable to find enough wet-nurses for the hundreds of babies in their care, they explored substitutes ranging from goats' milk and donkeys' milk to a new plant highly recommended as infant feed: the arrowroot, originating from Cuba. Alas, the powder did not live up to its promises, and after a dozen babies had died the experiment was halted.

These projects of the enlightened imagination reflected the nutritive value of food and its ability to nourish and increase productivity as well as happiness. As new knowledge accumulated about the value of foodstuffs rich in starch, the potato came out as Number One in starchy content and hence nutritional value (Earle 2017). But the 'quantum of happiness' which had been so proudly displayed as an icon fell pitifully and tragically short of the universalist aspirations of the European Enlightenment.

The story of breadfruit and potatoes tells us about the role played by the human body in the Enlightenment vision of a happy future, and invites comparison with how that vision has changed over time. Industrialization began with goods being mechanically produced on a mass scale, and later took over agricultural production as well. Machines as the embodiment of invested capital still depended on the labour performed by workers, who had to be nourished and kept healthy. Improvement of their conditions lay at the heart of the political struggles that accompanied the Industrial Revolution. The narrative of progress continued to gather speed and accompanied the rise of the consumer society. Progress never was the tide that lifted all boats, but it lifted many. Breadfruit remained an exotic item on the menu or on view in a botanical garden, while potatoes, together with rice and wheat, continue as staple foods for millions of people around the globe, although they are no longer associated with public happiness.

Today we have entered the stage of automation, of Industry 4.0 and the Internet of Things. Far fewer workers are needed and neither their nourishment nor public happiness are on the political agenda. Instead, governments are keen to support innovation, with digitalization accorded a high priority, as well as ensuring citizens have the digital skills they will need in the future. But the quest for happiness has not disappeared. It has just been privatized. From providing starch-rich food for workers, it has moved on to the endless possibilities of human enhancement in relation to the physiological functions of the human body, cognitive performance and mood enhancement – all at the service of individuals who can look forward to longevity. Predictive algorithms provide a new kind of digital nourishment consisting of recommendations for a healthy lifestyle. They offer digital feedback on how to attain, maintain and further enhance individual happiness, now reduced to a simple utility by the happiness business (Davies 2015). The services, therapies, apps and other products offered by the happiness industry are further mediated by councillors, coaches, therapists and self-appointed happiness experts (Cabanas and Illouz 2019). Instead of breadfruit or potatoes, algorithms now furnish us with our daily 'quantum of happiness', including making us walk the famous 10,000 steps.

Another striking difference compared to the world of the last two centuries is that we have become aware that planet Earth is finite and will soon have to sustain a population of 8 billion. The expansionist drive of the eighteenth century was focused on what was still to be discovered and taken into possession – plants, minerals, animals, people and land. The equivalent of Captain Cook's voyages today are the vessels equipped to descend to the deep ocean floors, surveying the terrain for minerals and new sources of energy. Spacecraft are designed to bring back minerals from the Moon. The curiosity about geographic discoveries has been replaced by curiosity about a digital mirror world with interactive avatars of everything that exists on the planet.

Meanwhile we are busy carving out our digital niche amid the fragility of the natural ecosystem. Instead of transporting breadfruit trees around the globe, we set up crop trusts, seed banks and agricultural laboratories to restore higher levels of crop and marine diversity, which has been rapidly altered by climate change (Stokstad 2019). A UN-backed report warns that 1 million plant and animal species are at risk of extinction, a dire threat to the ecosystems on which people, and not only animals and plants, depend. About 75 per cent of land and 66 per cent of ocean areas have already been significantly altered, and climate change is pushing a growing number of species closer to extinction (IPBES 2019). At the same time cities become the refuge of animals who are losing their natural habitat due to rapid urban expansion. Darwin is indeed 'coming to town' as the urban jungle drives evolution (Schilthuizen 2018). Agriculture is moving rapidly into the city too, with cultivation on the vertical surfaces of high-rise buildings. Artificial Intelligence is tasked with exploring the optimal conditions for growth and how to boost the yield. CRISPR, a new tool for editing the genomic inheritance, is altering plants in ways that would have been unimaginable to breeders of earlier centuries.

The fact that we call this progress shows that the narrative of progress, which grew out of Enlightenment imaginaries filled with ambition and folly, ideals and greed, is still with us. But the challenges and doubts are mounting. We have abandoned the effort to pursue public happiness and now leave it to each individual to find their own. But where is our shared purpose and our collective imaginary for a digitalized future?

Human enhancement, control and care

To understand the dominance achieved by the narrative of progress, one has to get to its normative core. The narrative presents progress as an inevitable force that originates from the deep-seated desire for improving the human condition.

It started with existential material conditions and moved along the pathways laid down by a combination of techno-logical invention, human will and planning foresight as practised during modernity. Now we have reached a stage where the focus of further improvement is increasingly on human enhancement. This finds expression in the yearning for a perfect body and a sharper mind, in the aspiration for a longer and healthier life, delaying the ageing process or, beyond that, entering the fantasy land of a transhumanism that will bring us closer to immortality.

The challenges to the narrative of progress arise from an increase in complexity that is at odds with the notion of a linear process. Improvement is still seen as desirable, but no longer as sufficient when we can no longer project it along a straight line leading from good to better without considering other changes and the unintended consequences of human action. In *The Collapse of Complex Societies*, Joseph Tainter, an archaeologist by training, describes how human societies develop increasingly complex structures. His main argument rests on the idea that social complexity is subject to dimin-ishing marginal returns. It costs ever more to produce ever smaller benefits and more resources are necessary just to keep complex societies going. Triggered by an outside event, they eventually collapse (Tainter 1988).

One need not subscribe to any of the grand narratives that relate the familiar rise and collapse of great civilizations. A growing number of scholars have begun to question the notion of 'collapse' as inviting a one-sided interpretation. When the experience of those who were most affected is included, the processes of ecological over-exposure and the final blows dealt by a disaster are better captured as the fragmentation of centralized power structures that results in the dispersal of a population to the surrounding environments, often liberating them from forced labour (Scheidel 2019; McAnany et al. 2009). There is agreement, however, that human societies evolved by generating complex internal structures. Archaeologists continue to find remnants in remote parts of the world of amazingly complex societies that disappeared thousands of years ago. Evidently, these societies were vulnerable and unsustainable, a message that resonates strongly with present-day concerns.

Given such long timescales, the narrative of progress is young. It served its purpose during modernity, but the question is whether it can be sustained in view of the levels of complexity that have arisen in today's highly interconnected global world. Its normative core is masked, as it is not capable of acknowledging the downsides of the unintended consequences that progress brings with it. Only by presenting itself as value-free and neutral can it sidestep the dark side of progress and its collateral damage. It cannot be held accountable as its credibility would suffer. This is why it has to continue to present itself as the only road to a shiny future that is open to all, insisting that opportunities only have to be seized and social progress will follow. As the saying goes, 'shit happens', but progress will continue.

But there is a hidden side to the narrative of progress, which it conveys only implicitly. It is the promise of control that would come with continuous improvement. The issue of control becomes obvious when control fails, when technological systems break down or when a system evidently fails to deliver what it had been designed to deliver. Implicitly, the narrative of progress is about extending the range of control. Of course, control is never absolute, nor can it ever be completely assured. In many instances it turns out to be an illusion. When control fails, accidents happen and a retrospective analysis is undertaken in view of preventing them from reoccurring in the future. In this respect, progress has undoubtedly been achieved. Accident prevention, at least in most highly industrialized countries, can proudly point to the gains that have been made. Whether it is control over machines and human–machine interfaces, or over avalanches, flooding and other natural disasters, or in the medical domain – everywhere safeguards have been put in place with the requisite preventive measures, protocols, checklists and training of personnel. Accidents still happen, but accident prevention is writ large. On this score, the narrative of progress has been fully vindicated.

But control poses a bigger challenge when it relates to socio-technical systems. The first time the limits came into full public view was with the radioactive waste left behind by the military production of nuclear bombs, followed by the civil production of nuclear energy. The half-life of radioactive substances is outside human control, which makes it necessary to store the waste in places considered to be safe. Some of these sites have been outfitted with special warning signs in the hope that future generations and even other civilizations will be able to read and understand them. They are awesome monuments buried in deserted and well-secured places, as are the containment structures erected in Japan after the Fukushima accident. The Three Mile Island accident in 1979 involved a partial meltdown of one of the reactors with subsequent radiation leak. It caused dismay worldwide, fuelled the anti-nuclear movement and instigated serious inquiries into the nature of complex systems and how to deal with them. 'Human error' can no longer be blamed when complex systems threaten to collapse. Loosely coupled systems, for instance, fare much better when something goes wrong than tightly coupled ones (Perrow 1984).

The linear logic underlying the narrative of progress does not fit with the non-linearity that is a constituent feature of the dynamics of complex systems. The obvious mismatch evokes concerns about the scope and depth of human ignorance when facing genuine uncertainty. The linear assumption underlying a control-oriented technocratic order calls for a politics of uncertainty that could pave the way for a culture of care (Scoones and Stirling 2020). Others have called upon policy-makers to adopt a set of 'technologies of humility' with which to assess the limits of scientific knowledge and to work out how best to act under irredeemable uncertainty (Jasanoff 2007). The most recent example of such a situation was the COVID-19 pandemic and the helplessness with which many governments reacted to it. It laid bare the global interdependency between health and the economy and their vulnerabilities in a volatile geopolitical context. The narrative of progress has little to offer when nobody knows how to cope with the long-term consequences of the pandemic either at the local or the global scale (Delanty et al. 2021).

More such predicaments await us in the future. Epidemiologists warn about new epidemics. In many cases data from the past can be a reliable guide for the future, helping us to prepare for the known unknowns, such as earthquakes and the outbreak of epidemics. They have a high probability of reoccurrence, but it is not known when.

The unknown unknowns, however, are those very rare events whose impact is often huge. They are unknown in the sense that they have not happened before, rendering their prediction impossible. But even if accurate data can warn us ahead of time, such warnings may fall into an institutional void if they are not heeded or preparedness is lacking.

Preparedness begins in the mind, but another hindrance appears for the narrative of progress. If facing complex systems in the present is already a big challenge, it becomes nearly impossible to even think of being in control when faced with the complexity of the future. Research is an inherently uncertain process, and innovation is likewise beset by many uncertainties regarding its outcome. The more techno-scientific options become available, the more novel combinations arise and only a very small segment of what is feasible will be realized. To take just one example: we are at the edge of breakthroughs in quantum computing, which has moved from being mostly a theoretical curiosity to potentially solving problems that would take impossibly long on a classical computer. We have no clue how quantum computers will affect society in the future nor how society will deploy them.

The fact that the future remains inherently uncertain is nowhere more evident than in biological evolution. It has no telos built into it. Some organisms, like bacteria, have persisted without much change for millennia, others have gone extinct and new speciation occurred. Natural selection remains a crucial mechanism, with random as well as targeted mutations, genetic drift, epigenetic forces and other more fine-tuned processes operating at different levels of life. Biological evolutionary processes are not the same as those underlying cultural evolution, but there is growing recognition of the need to extend the biological evolutionary framework to include the environment and ultimately society (Laubichler and Renn 2015). Evolution has brought forth some of the most stable mechanisms capable of preserving certain features against change. At the same time, it continues to generate novelty, experimenting playfully with new possibilities to see what works under which conditions.

Within such an evolutionary framework it makes little sense to speak of progress. And yet, in the long history of biological evolution that proceeds through trial and error, the unlikely event of cultural evolution occurred. The human species, itself the result of the evolution of several lineages of precursor species, began to develop language, consciousness and cognitive abilities that eventually created modern science and technology. Cultural evolution has since overtaken biological evolution. It has brought us to the point of our development as a species where we are able to create computational machines that we use to predict future events.

In their analysis of the major transitions in biological evolution, John Maynard Smith and Eörs Szathmáry identified the features common to the various stages, from the first replicating molecules to multicellular organisms on to eusociality and the acquisition of language. Smaller entities come together to form larger ones, leading to differentiation. But the most salient feature of every major transition in evolution is the emergence of new ways of transmitting information, whether it be transmitting DNA proteins, cell heredity, epigenetics or universal grammar in language. These changes are highly significant as they lead to a new stage in evolution. This is not simply 'progress' in the sense of improving the conditions for survival or better reproductive performance. The patterns of evolutionary transition clearly move towards an increase in complexity (Maynard Smith and Szathmáry 1995).

New ways of transmitting information are the hallmark of the computational power behind the latest technological feats that have become almost daily routine. During the last days of 2020 China's Chang'e-5 mission succeeded in scooping up lunar rock and dust from a previously unexplored region of the Moon's near side and returning them to Earth. The latest algorithmic programme by the AlphaFold AI system is able to predict the fiendish problems of protein's 3D structures with greater accuracy and speed than experiments can achieve. As mentioned before, supercomputers have been deployed to reduce the 1 billion possibly therapeutic molecules to less than a few thousand as part of the work of the COVID-19 High Performance Consortium. And, most recently, *Perseverance* transmits intriguing information directly from Mars about the surface and other conditions there.

Faced with the inherent uncertainty of the future, however,

no algorithm is able to predict the unpredictable. We have no control over the atmospheric or ocean circulation cycles that drive climate change. And while the narrative of progress keeps whispering into our ear that eventually we will be able to control this last frontier, the future remains uncertain. Simulations of complex systems with real data help us to extend our view. But we can no longer hide from the fact that we are not in control. Complex systems are largely self-organizing. This renders the defence of the beleaguered narrative of progress futile when it insists that humanity is wealthier and healthier than ever before. Graphs and curves that register improvements do not match people's feelings about whether they are better or worse off. The gap between the subjective perspective and that of statistical averages persists.

Hans Rosling was a great communicator of facts and figures, and he too tried to convince his audience that the world is better off today than it used to be. Humanity has achieved progress over the past century and a half, he argued, despite the setbacks and the horrors of the twentieth century. Progress is real, and is not a question of optimism versus pessimism. 'Factfulness' for him consisted in acknowledging that progress had two sides, 'bad and better'. What he meant was that we have to include the downsides – the unwanted effects that inevitably accompany things getting better – in our overall assessment of what has improved and what kind of progress has been achieved. With Scandinavian modesty and forthrightness, he presented himself as a possibilist, as someone who neither hopes nor fears without reason (Rosling Rönnlund, Rosling and Rosling 2018).

However, before we consign the narrative of progress to the dustbin of history, remnants of it may be worth saving, ready to be repurposed if we want to make something new out of them. The narrative is broken because it cannot change tack and adopt a holistic approach in facing the challenges ahead. It cannot preserve a hopeful message of continued improvement when seemingly insuperable problems block its way. Nor can it present a credible outline of how to approach the problems that arise from dealing with complex systems if it remains stuck on its old message.

One remnant warranting further analysis and re-assessment

is the demand side of the narrative, the inveterate human desire for a better life and for continued improvement. The demand is inexhaustible, as human wishes and desires are infinite. The more goods and pleasures are delivered, the more the imagination is enticed to crave new ones. Capitalism in its highly adaptable forms has flourished by exploiting this insatiability of human wants and needs. Digital technologies are ideally suited to keeping the spiral of desire moving upward, creating new points of entry and exit between the virtual and the real worlds, between the play of the imagination and the work that keeps the economy going.

What began with satisfying the most basic existential needs has developed into the wonderland of consumerism, offering choices among an endless supply of digital gadgets, from the useful to the frivolous, the mindless to the educational. But we can also opt, as the International Panel on Social Progress did, for a vision of a better society and engage in serious discussion on how to reach it. Such a goal is a huge undertaking, weaving together the technological strands with those of social justice, undergirded by the belief that such a society is possible (Fleurbaey et al. 2018).

Another boost to the waning belief in progress, but with a decisive twist, comes from the scientific community, which always held on to its own definition of progress, assessed solely according to scientific criteria. The narrative of progress that circulated in society was welcomed because it supported science, but otherwise considered of little relevance. However, recent developments have forced a remarkable rethinking. Human control over life and living organisms is rapidly expanding through new techniques such as CRISPR that make precise gene editing possible. Tissue regeneration in mammals becomes possible by reprogramming cells to a 'younger' state in which they are able to repair or replace damaged tissue. These and many other interventions demonstrate the amazing advances made by science and technology in actively changing Nature, both inside us and outside.

Science has definitively arrived in the midst of society, and society wants to be included in discussing what science is now capable of doing. Users, patients, environmentalists, as well as vested interest groups, governments and courts – all want their voices to be heard. Society talks back to science, and science is obliged to listen. Moreover, technocratic expertise alone is no longer accepted, nor sufficient to legitimate decision-making. Seemingly endless deliberations and struggles over regulatory, legal, financial and other control mechanisms may seem tiresome, but there is no way around them. Designing systems of monitoring and control over what cannot be completely controlled demands a greater willingness to listen to those who will be affected by decisions in which they do not take part. This requires more than just following ethical guidelines for research, indispensable as they are. A new culture seems to be in the making. Whether it will be a culture of accountability or a culture of care remains to be seen.

Similar discussions are ongoing over the monitoring of digital technologies, focused on how much autonomy can be transferred to machines and how to prevent them gaining control over humans. Among the issues is that of accountability, evoked every time a major technological glitch happens or a disaster occurs. The entanglement of technical faults and human error muddies the issue of who is or should be in control and hence held accountable, making it challenging to find adequate, enforceable regulatory mechanisms. But the more a narrative of novelty and innovation edges into the void previously filled by the narrative of progress, the more such issues must be faced and satisfying responses found.

Science and technology have opened up an enormous number of possibilities that lie in the future. The theoretical biologist Stuart Kauffman introduced the concept of the adjacent possible. He asked how evolution had worked with big data, taking as an example the enormous number of possible combinations and permutations of proteins needed to come up with the protein formula that constitutes life (Kauffman 1995). If the biosphere continues to expand in order to increase diversity, how does it do so and at what speed? Among the exponential possibilities, the further we go into the future, the more possible technologies can be invented, the more possible thoughts can be thought, and the subset of the actual becomes smaller than a fraction of the sum of what is realizable.

The adjacent possible begins at the edge when we look

from one set of data into neighbouring data sets. The adjacent possible is the space to explore if we want to find the new, starting at the edge of what is known and looking to uncover the next new item, idea or thing (Loreto 2018). But it does not pretend to tell us how much control we can have over the future. This is a preposterous proposal as the future cannot be controlled. Yet prediction challenges it, since it is intent on extending control. Predictive algorithms should help us to be better prepared, but, as we have seen, they can also mislead us into attributing an unwarranted agency to them.

Perhaps we no longer need a narrative of progress like the one conceived during the Enlightenment. It has served us well in helping to bring about the progress that has undoubtedly occurred. But it also has failed many who believed in the promises that have not been fulfilled. Above all, the narrative of progress has so far not shown sufficient flexibility to adjust to the major changes and future challenges awaiting us in the twenty-first century. If it is to survive, it will have to appeal to a society whose pluralism and aspirations it must incorporate. Rethinking what progress means in the digital age confronts us with the uncertainty of where our interaction with digital machines will lead. Reality has forced us to abandon the fantasy of human dominance and our complete control over what we do and plan. It invites us to cultivate instead the ability to embrace uncertainty.

The quest for public happiness through the supply of breadfruit and potatoes turned sour and never realized its goal. In retrospect, it may have been one of those impossible dreams of the Enlightenment that also demonstrates the limitations of a technological fix. But this should not prevent us from trying again to pursue the public good, this time on a planetary scale. The narrative of progress has accompanied us on the road humanity has travelled so far. Undoubtedly, much has been achieved. It has nurtured the illusion of our control over nature and led to a human-built, technological world at the expense of the natural environment. We are now obliged to acknowledge its dark sides as well. But the desire to improve the human condition persists. Any new narrative will have to incorporate accountability and the limits of human control while allowing ample space for redefining its extension. Control will need to be balanced by care. We learn by doing, by practice and usage; in other words, by experimenting. Last, but not least, while control includes and will increasingly depend on prediction, prediction itself also needs to be controlled.

4

Future Needs Wisdom

Why wisdom is needed

There are moments when a phrase comes up that settles in one's mind. It happened to me on the way back from a conference on the future of AI when I decided to write a haiku. This unfamiliar urge was triggered by the haikus that an AI had composed for the conference and which I found, well, somewhat bland. Readers can judge for themselves, as this is what the AI charged with the task of composing a haiku for each of the two groups, named 'Artificial' and 'Intelligence', had to say:

Group 1: Artificial
finally new world
artificial delegates
before the mankind

Group 2: Intelligence
swift reality
where intelligence captures
at the perfect risk

In my haiku, the first ever I wrote, I combined the terms 'artificial' and 'intelligence' that had been given to the two groups and their haikus and came up with:

Artificial Intelligence
human company
invisible algorithms
future needs wisdom

Given the capability of the human mind to make sense of patterns that have no sense and to impute meaning to words without meaning, we could delve further into what the AI spurted out. We could guess endlessly about what the artificial delegates have to say when invited to speak before 'the mankind', or what intelligence captures and which are the perfect risks this entails. But this is not the point. Following the last line of my haiku I simply want to argue for why the future needs wisdom.

As a statement it might be contested. Where is the place for wisdom in contemporary societies, let alone the feeling that it is needed? Wisdom appears as a residue from a premodern world that originated long before knowledge was written down or became codified and even longer before science arrived as the main producer of new knowledge. We associate wisdom with oral traditions in which knowledge was preserved and passed on by the elders of the tribe to the younger generation. Or it might have been preserved in timeless proverbs or in obscure manuscripts. We still recognize moments when we receive wise advice from others in undecided and difficult situations, the wisdom generously shared by wise women and men with rich and diverse experience that has sharpened their critical judgement. They are not bound by interests of their own or of those close to them. It is certainly good to have them around, but there is no reason to expect that their wise council will be on hand when needed.

The wisdom I appeal to in my haiku is of a different kind. It is a set of attitudes and practices that pervades institutions and organizations, percolating through all levels of a system. It is an *ethos* of shared practices that is continuously cultivated, adjusted and refined in order to be exercised as and when the situation demands. It resembles cathedral thinking in being capable of bringing together individual talents and skills, in order to carry forward an open-ended effort that requires everyone to unite behind it. This is why it is needed as an institutional feature, a recognizable property of a system. It excels in being able to look forward and backward at the same time and in establishing a balance between the urgency of the moment and the long-term view.

Looking backwards means accessing the richness of

experience that humanity has accumulated in the past, selectively stored and retained in physical artifacts, rituals and texts of various kinds, written in a multitude of languages and genres and dispersed around the globe. This heritage is valued for its transmission of what we call culture, passing on the spectrum of the ways in which human creativity has expressed itself. Similar to what happens in biological evolution, where the transmission of information encoded in genes and their regulatory networks is essential to ensure the existence of future life, culture forms an essential part of where we come from and who we are. Just as Nature has found a way of being ultra-conservative in preserving some of life's essential features across species and over time while experimenting with bringing forth novelties, cultural heritage builds on tradition, but stands for innovation, when we succeed in recombining elements from the past with the creative impulses of the present.

Andrea Nanetti has drawn attention to the necessity of aligning cultural heritage with AI and machine learning algorithms to decode and encode the treasure of human experience in a genuinely transcultural and multilingual approach. This entails much more than scanning old books and manuscripts, more than the toil of translating and digitalizing languages that nobody speaks today (Nanetti 2021). The humanities, a vital domain of knowledge, are overwhelmingly researching and reworking the human past. Much of it has been lost forever, but the application of digital technologies can generate amazing results in retrieving the deeper layers to be found in unexpected places. Looking not only backwards but also forwards highlights a problem that is continuously faced by archivists, librarians and other scholars. This is the decision about what to discard and what to retain, given the sheer amount of material not only from the past but also that which accumulates in the present. Wisdom consists in linking the past with the future, advising what to do in the present. It is about rendering knowledge retrievable for questions that have not yet been asked. The humanities, just like basic research conducted in other scientific domains, produce knowledge that may seem utterly useless at the time, but wisdom consists in facilitating the emergence of its future usefulness.

This is what culture is about. The question of how to create and organize knowledge about the treasure trove of human experience in order to keep it open for future use has led Umberto Eco to retrace the major stages in doing so. The path leads from the notion of the tree of knowledge based on an orderly but closed classification system, to that of a virtual encyclopaedia that is open and interconnected with all other virtual encyclopaedias. In concluding he writes: 'If cultures survive, one reason is that they have succeeded in reducing the weight of their encyclopedic baggage by placing so many notions in abeyance, thus guaranteeing their members a sort of vaccination against the Vertigo of the Labyrinth' (Eco 2014: 93; quoted in Nanetti 2021).

Here it is again, the vertigo of the labyrinth of which I was at first only dimly aware on my journey into digi-land, the overload of information that squeezes the temporal experience of the present and easily leads to exhaustion. It points to what is at stake, and this is where wisdom enters: What is to be kept in abeyance, latent and dormant, but ready to be retrieved and recombined with other knowledge to face the problems of the future, a future about which we know nothing? Which criteria should we use to decide what to keep, and how can our preparedness be enhanced so as to recognize the significance of what has been found by not looking for it, thanks to serendipity, that potent ally of science?

Nobody, not even the most fervent AI enthusiast, would claim that an AI exists that is able to decide which knowledge is to be kept in abeyance to ensure the future resilience of a culture or society. There is no 'wise AI'. Instead we have the 'smartness' that is designed into every algorithm and every digital device and system. To be smart means to be smarter than others in order not to be outsmarted by them. It stands for efficiency, for delivering on predetermined goals. Everything has to be 'smart' now, from cities to our homes and how we work. A smart home will save on your electricity bill and a smart city will be run smoothly and efficiently, mastering traffic flows, the reduction of CO_2 emissions, and an environmentally friendly recycling system. A smart working place is a digitalized environment that is well connected with other digitalized environments and working places. Smart AIs are designed to fulfil a given purpose, a function that meets well-defined and quantified criteria, as this is how algorithms work. It is this feature that Brian Arthur has in mind when suggesting that we should think of algorithms as verbs: tell them 'do this', 'store that' or 'retrieve this' and they act by following the instruction they are given (Arthur 2018).

This is why it is so difficult for algorithms to deal with ambiguity, with the half-tones and innuendos that pervade human conversation, the smiles or the silences that are non-verbal communication cues meaning different things in different cultural contexts. The reason why future needs wisdom is that we will have to go beyond the well-defined purposes of an AI, to delve into a variety of contexts that are subject to continuous change, to take meaning into account where an algorithm acting as a verb knows nothing beyond the purpose for which it has been designed. In other words, algorithms will have to acquire culture, a sense of how to maintain a balance between what is to be discarded and what is to be kept in abeyance, if they are to serve as a vaccine against the vertigo of the labyrinth to which they themselves keep adding new features.

Future needs wisdom means to practise an ethos that consists in finding ways of tapping into the resources that the past and the present offer for guiding behaviour, widening the horizon and helping to design the new institutions we will need in the future, capable of providing answers for tomorrow's problems. Wisdom will reside in mechanisms that can bridge the gap between the individual and the community, between the individual and what happens at the level at which complex systems operate. Wisdom is needed to respond to what is not yet foreseeable. It is not about 'solutions', though it may help bring them about. It is the opposite of a quick fix, be it technological or of any other kind.

Awareness about the problems and issues where wisdom is needed is growing. They begin with the sustainability crisis of the digital Anthropocene. Other concerns revolve around the one big question: what uses or misuses will humanity make of the computational power now at our disposal? Ongoing discussions focus on the further loss of privacy and the many possibilities of surveillance, the threats to liberal democracy and how to curtail the circulation of ever more fake news. Many people are worried about the future of work. We boast about the digital prowess we have achieved, but can barely conceal our anxieties about the redefinition of our identities and of what it means to be human. Part of the discussion concerns the actions and measures to be taken, including the calls for ethical oversight and for the design of AI systems that are beneficial and responsible, aligned with values we can barely agree upon. We know about the urgent need for more and better regulation of AI and of the big corporations that own them, yet the political will to act robustly is still lacking. In all of this, wisdom is needed; the kind of wisdom that acknowledges the limitations of digital technologies and guards against the illusion of control.

In facing these challenges, we are thrown back on our own resources, tasked with inventing the institutions that will be needed to mobilize and manage the major transitions ahead. This holds also for the other issues that preoccupy us, including how to ward off the threats posed to liberal democracies. These were laid bare by the shocking events at the US Capitol on 6 January 2021, at the heart of the world's leading power, while the whole world watched. And what does it mean for tech regulation and democracy itself if it is left to a large corporation to curb lies and incendiary speech online, and will such efforts to curb abuse inadvertently strengthen state surveillance? What else can and must be done to keep discrimination at bay when predictive algorithms are trained on data that is already biased? Calls for an ethical AI are ubiquitous, but where do we actually stand if the data continues to neglect local contexts and diversity?

Pathologies of AI: ethics is more than a checklist

The generally accepted premise is that AI is not intended to replace humans but to complement them. It should enhance human capabilities. When the chess programme Deep Blue beat the world's best chess player, Garry Kasparov, many thought this would be the end of the game. But that is not what happened. The chess programmes now available on laptops and smartphones help people improve their playing skills. But the ambition for AI is to go further, towards becoming what is called Artificial General Intelligence (AGI). This would put AI at the same level as human intelligence, which is characterized by its generality. While that is unlikely to happen anytime soon, if ever, the widespread adoption of neural networks that has enabled Deep Learning and other sophisticated computational procedures has initiated an important new step.

It is a transition period of amazing advances and of worries. The increased dependence on AI that is happening across society has also brought an increase in cyberhacking and the risk of AI breakdowns of various kinds. AI not only changes what we know, but also how we know and which values will prevail. This gives rise to many tensions, and desperate efforts are under way to not let the development of AI get out of control. Notwithstanding the appeal of the ubiquitous calls for ethical guidelines and for designing an AI that is responsible, beneficial and aligned with human values, a host of technical obstacles and, more importantly, a lack of concrete guidelines on how to move towards actionable goals, characterizes the actual situation. There is an overestimation of the technology and an underestimation of the social contexts in which AI is operating, whose interests it serves, and in which political and economic circumstances.

Some authors, like Edward Ashford Lee, describe the abuses and misuses of AI as pathologies (Lee 2020). A pathology presupposes some kind of reference to what is considered normality, as Georges Canguilhem's seminal work *Le normal et le pathologique* demonstrated for institutionalized medical knowledge (Canguilhem 1966). If already medicine and biology struggle with the definition of normality, this is even more the case for the relatively young and not firmly institutionalized field of AI. Nevertheless, if only as an analogy, the characterization of the many flaws of AI as pathologies carries an additional message. Even though some pathologies will persist and will certainly mutate, the message is that many can be overcome. The analogy with disease also makes it clear that the standards that define normality and pathology are not immutable. It is, for instance, up to us whether to accept as 'normal' the hate speak and fake news generated by bots and widely circulating on the internet, or to treat it as a pathology for which an educational vaccine has yet to be developed.

The susceptibility of AI to bias is one of its most visible, acknowledged and harmful features. Bias is not so much deliberately designed as surreptitiously introduced through data that already carries a biased load. All of us are subject to bias and prejudices, and we often re-enact them unconsciously if we are not explicitly made aware of them. The vast range of cognitive biases that affect how the human mind operates has been laid bare by the ground-breaking work of Daniel Kahneman and Amos Tversky (Kahneman and Tversky 1979; Kahneman 2011). Moving from the individual to a societal level, it is not surprising that we find widespread prejudice and discriminatory practices reproduced. One particularly egregious example is the development of facial recognition tools that have much higher error rates for black faces, yet so far little has been achieved to rectify this.

Technologies are thus intrinsically intertwined with conscious or unconscious bias since they reflect existing inequalities and discriminatory practices in society. Biases exist everywhere, including in phenomena that can be measured and those that remain largely invisible or even unconscious. Misuse and abuse are exacerbated by poor data quality collected from untrustworthy sources or processed in sloppy ways. 'Trash in' leads to 'trash out', as is well known. In the end, mathematically consistent algorithms that are based on incorrect or poor data can do more harm than a compassionate human who is prone to bias and error, as the former cannot easily be corrected or counteracted.

By delegating ever more decisions and tasks to AI, human responsibility is diluted. This is exemplified by well-known cases from the United States, where police profiling and court sentencing are already highly dependent on predictive algorithms, as is access to housing and to insurance (Fry 2018). These cases underline the urgent need for regulation. Distributed agency must be taken into account in systems where the extent and kind of control each actor has differs greatly and their actions and intentions are difficult to attribute. One solution is to hold all agents in the system, such as a company, responsible.

The AI industry has not been indifferent to these harms, and recognizes the distorted outcomes that algorithms can produce. One of the first attempts to highlight the issue of algorithmic bias dates back to 2014, when a group of AI researchers from academia and industry joined 'thought leaders' from various backgrounds to found the Beneficial AI Movement and the Future of Life Institute (Tegmark 2017). More conferences and declarations followed, but no easy solutions have since emerged to ensure that AI technology is indeed 'beneficial'. The technical problems turned out to be not negligible and they continue to challenge AI designers and engineers. The effort to institutionalize a regime of governance for AI definitely has to continue (Russell 2019).

Less in public view are the alerts that Google and Microsoft send to their investors. They warn that the algorithms embedded in their products and services could give rise to ethical, legal and technical challenges that may negatively affect revenues and operating results. Microsoft goes even further, admitting that its algorithms may be flawed and its data sets insufficient or have built-in bias. The warning is clearly spelled out: such deficiencies could undermine AI applications in ways which may lead to legal liability and brand or reputational harm.

A concrete example of such a deficiency occurred during the summer of 2020 in the UK, after A level exams had to be cancelled due to the COVID-19 pandemic. The Education Secretary Gavin Williamson introduced a new procedure for assessing A level grades. An algorithm would be deployed to review the previous assessments of students by their teachers in order to prevent any potential bias and to avoid grade inflation. Introduced as a 'moderation system' the algorithm led to a reduction of almost 40 per cent of the grades given by teachers. It disadvantaged high-performing students from poorly performing schools and all students from schools whose results were improving. These disturbing results were obtained because the algorithm included factors that had little to do with the students' performance and disproportion-ately affected students in schools located in disadvantaged areas. A public outcry followed as students' university applications were seen as being unjustly rejected. In the end, the algorithmic 'moderation' was cancelled and the teacher-awarded grades were restored.

This incident may represent nothing more than a minor, pandemic-related algorithmic hiccup. But it highlights deeper issues related to the increasing deployment of algorithms and machine learning procedures in various policy areas. As the A level example shows, decisions based on algorithmic results can have substantial implications for those affected. Such decisions raise issues of political accountability and legal compliance. They also create distrust. Algorithms, as Brian Arthur is fond of saying, should be regarded as verbs. They do things. And they can also do them badly.

What at first sight looks like a technical problem is therefore first met by offering a technical solution called 'explainability'. This refers to methods and techniques that allow the results and solutions offered by an AI to be 'explained' in a way that can be understood by humans. It amounts to prying open the black box in machine learning, where often even their designers cannot explain how an algorithm arrived at a decision. To avoid outcomes that are patently unjust or considered biased or illegitimate, the hope is that when it has to be 'explained' to users how machine learning systems are designed, this will improve, if not their real understanding, then at least their grasp of the 'black box' nature of algorithm-based decision-making. Although 'explainability' is treated as a technical problem by the AI community, it goes far beyond the technicalities that are involved. At a technical level, at least, it has induced the big tech companies to quickly invent a technical solution – installing an AI tasked with doing the required 'explaining'.

The real problem is rather a fundamental incompatibility between the logic of algorithms and that of policy-making. As Diane Coyle and Adrian Weller describe in detail, a machine learning system is set up to optimize an objective function in the system which is the functional equivalent of human intention. Therefore, it must be made explicit what the system is expected to achieve and which objectives, values and political choices should be incorporated in the design. This is in stark contrast to most policy-making, which typically relies on what are called constructive ambiguities to pursue shared objectives. These are often the result of political compromises, considered necessary in order to achieve consensus. Political and policy decisions usually include trade-offs between multiple, often incommensurable, aims and interests. The algorithms in machine learning systems, by contrast, are utilitarian maximizers of what is ultimately a single quantity based on explicitly weighted decision criteria. They do not tolerate ambiguity.

Coyle and Weller speculate what is likely to happen if the demand for explanations of algorithmic decision-making systems prevails. There is an underlying tension that has to be brought into the open. The tension is between human decision-making and machine-supported decision-making, which can only be resolved by forcing greater clarity about the choices and trade-offs that underlie the former and which are left implicit or ignored in the latter. It will be interesting to see how the political and policy-making system reacts to such a challenge. Until digital machines have learned to observe and incorporate human contradictions, errors and ambiguities – a prospect which is far in the future – there needs to be space for openly discussing conflicting objectives and existing contradictions. They will need to be resolved in an unequivocal way, because only then are they quantifiable and capable of being incorporated into machine learning (Coyle and Weller 2020).

An example from the Austrian Labour Market Service (AMS) illustrates the dilemma. The AMS is a well-run public agency whose portfolio includes supporting unemployed persons to find a job by matching offers with their personal profile. It also offers additional training and other services. When the leadership of the agency decided to introduce algorithms into their matching services, the stated goal was to spur efficiency by alleviating the workload, speeding up the process and cutting costs. An algorithm was designed to divide the job seekers into three groups, depending on criteria such as age, sex, qualifications, previous employment record, duration of unemployment and the like. Based on these data and the known outcomes of finding a job in the past, the algorithm predicts which configurations of characteristics will increase or lower the likelihood of finding employment. The logic behind introducing the algorithm was very simple. An automated decision-support system helps the agency staff concentrate their efforts on the two groups whose chances of finding employment within a reasonable period of time are higher. The third group with the lowest chances would also be attended to, but with much less time and effort put in by the agency staff.

When the introduction of the algorithm became known through the media, a public outcry followed. The agency defended itself by arguing that its limited resources would be focused on supporting those who were most likely to benefit as early as possible. But the public anger fixed on what was seen as the overt discrimination against the most vulnerable group, the job seekers whose chances of finding employment were low. The agency argued that people in this group would continue to receive support, only somewhat less. Its critics were unimpressed. The problem, as they saw it, was that the data fed into the algorithm and the weighing of the different factors that had established the three groups reflected the status quo. In practical terms, this would deny a woman who had to take care of her elderly parents or disabled husband the same chances as another woman of the same age and similar track record but without these responsibilities. Those who were hardest hit would thus remain locked in the situation they were already in.

On the surface, the classification of job seekers into three groups according to their chances of finding employment reflects their 'objective' chances, which is the basis for treating them differently. This is the old dilemma known as 'triage' on the battlefield. In the management literature it is called strengthening existing strengths. But the algorithm has no answer for what to do with the weakest and most vulnerable group. The dilemma persists, even if the ultimate decision is made by a compassionate staff member. Once a semi-automated decision-support system based on criteria of greater efficiency is in place, it will often function like a fully automated system as humans start to rely on it. If similar informal practices had existed before, they now become automated. Paradoxically, the AMS algorithm was not hiding anything. It fully met one of the criteria most frequently invoked in ethical guidelines for AI: full transparency.

This case is another reminder that what an algorithm does, how it works and what effects it produces are never merely technical issues. Algorithms cannot be separated from the social contexts in which they are embedded. In the case of an organization, whether an unemployment agency or the police, a court or a hospital, a traffic system or a bank issuing loans, if an AI is used to support its decision-making system, then the culture of the organization matters. Inevitably, that culture will influence the effects the AI produces which in turn will affect the human operators who are part of the overall organization in terms of how they use the machines and what the machines do over time to their judgements and decision-making.

The debate over whether or to what extent AI can be made to conform to ethical principles dates back to its very beginnings (Wiener 1960). Since then, there has been no lack of calling for ethical guidelines. All major national and international bodies that deal even remotely with digitalization have set up their own guidelines, and so have most governments and industries. What is not sufficiently recognized in all these efforts is the fact that with AI a new form of agency has been introduced. Clearly, an algorithm is not completely autonomous, but who is to be held accountable for its operations? While the focus of the European Commission is on trustworthy AI, the European Parliament has set up its own AI4People forum to help orient the industry towards what it calls a good society. The US National Science Foundation has put its weight behind AI fairness. In March 2019 Google formed an Ethics Board that was later dissolved amid controversy over possible conflicts of interest. Facebook has invested $7.5 million in a newly established centre for ethics and AI at the Technical University Munich. The list could easily be extended.

A research group at ETH Zurich decided to analyse eighty-four ethical guideline documents, written by enterprises, expert groups, governments and international organizations around the globe. About half of the declarations came from the private sector, the other half from public entities. To the research group's surprise, there was no single ethical principle that was mentioned in every document. Eleven principles were distributed in frequency across the entire range, and five key points figured prominently in more than half of the documents. These were transparency, justice, fairness, prevention of harm, and responsibility in conjunction with data protection. Despite this apparent convergence there was little overlap or agreement on what exactly was subsumed under each of these key points, nor on how to attain them (Jobin, Ienca and Vayena 2019).

Notwithstanding this fragmentation and divergence, the concept of transparency is the clear frontrunner in the quest for ethical AI. It reflects the concern to know what is going on behind the walls that shield the work of algorithms, how and for what purposes they are designed, and what limitations or hidden features the technology has. The more algorithms act as invisible facilitators between us and the environments with which we interact, the more important it becomes to lift the veil of secrecy, as invisibility enhances the power of AI. But as the example of classifying the job seekers shows, the performative power of an algorithm can be further enhanced by making it transparent. The complexities of real-life situations in a widening range of applications are an intractable mixture of the social and the technical. Even if we do get to peek inside the black box, it does not easily yield its intricate workings. Redesigning a computer architecture can be fiendishly difficult, as opaque layers have been built on top of more opaque layers. Some of the difficulties are simply due to complex code that is hard to understand even for insiders, while others are related to encryptions and safety features built in to ward off malware.

There continues to be an immense stream of proposals to strengthen the ethical backbone of AI. However, such accountability needs to be anchored on a solid legal basis, which in many cases does not yet exist. Auditing algorithms is among other proposals considered to be useful, but again, the rules to carry out such audits still need to be written. There is also the open question of what incentives there are for the industry to cooperate. Activists for an ethical AI agree that the industry must be induced to align with ethical guidelines, but warn that it should not be left to design the ethics it wants. Thus, it seems that everyone knows what should be done, but nobody is in charge of getting it implemented. Ethics continues to be evoked as a universal mantra but there are far too few moves towards norms and regulations that are actionable.

The 2020 annual Berlin Debate on Science and Science Policy was devoted to the changes to the research system as we know it that come with the rapid takeover by Artificial Intelligence. Participants were intrigued by the proliferation of AI ethics guidelines. They questioned how these general principles could be applied in different contexts. What is relevant and for whom? What needs to be prioritized for AI? How can these soft guidelines be translated into concrete and contextualized actions for the people designing and deploying AI? And how can the very different conceptions of ethics, fairness, social justice and other values that need to be put into context be spelled out in more detail?

Participants argued that it was time to stop the proliferation of AI guidelines and find ways in which institutions could set parameters in specific contexts, detailing what it actually means to be transparent, socially just or protective of privacy. They were also reminded that building ethics and fairness into machine learning models first requires a normative definition that can be translated into a mathematical function and then used as a constraint in some optimization problem. Barely any research on positive approaches to building ethics into AI exists, as opposed to pointing out what needs to be avoided. A positive approach would necessitate that machine learning models truly take into account local contextual definitions of what is considered to be ethical and how to conform to laws and regulations (Berlin Debate 2020).

Thus, ethics is not a checklist, where ticking a box absolves a company or organization from further responsibility. We are still far away from moving beyond rhetoric and good intentions. When discussing concrete cases the importance of taking context into account emerges time and again. Context matters and the discourse on ethics and AI might benefit from widening its perspective to look at other areas where the issue of aligning controversial research with ethical standards has been going on for some time.

Experience from the biomedical field suggests that there needs to be less reliance on ethical expertise and more attention given to representing those who will be directly affected. While building trust and ensuring transparency are essential in the pursuit of ethical science, so are inclusive participation and openness towards uncertainty, as opposed to distinguishing between a predefined 'is' and 'ought'. In the end, whether the perils and promises of application have to do with facial recognition, research on chimeric embryos, mass surveillance through contact tracing for COVID-19, or driving malaria resistance through the mosquito population, neither the illusion of techno-scientific control nor issuing new versions of ethical guidelines will protect us. This will only come by embracing our shared responsibility for human and non-human futures (Franklin 2019).

Wisdom refuses to be tied down to definitions, even if it may be advisable to have some common agreed-upon standards for ethical behaviour or a responsible AI. Instead, wisdom infiltrates local contexts and adapts itself to situational requirements and constraints. In doing so, it listens to the voices of those most affected and attempts to speak up for them, against the rigidities of rules and those who make and enforce them.

The enduring lure of the Leviathan

Our bodies may represent data points in digital space, to be tracked and their movements further processed, but at the same time we remain grounded in the social world where we move, eat, love, work, fight and age. This dual existence has given rise to serious concerns about surveillance and the further erosion of privacy. One of the most rapidly advancing fields in machine learning is facial recognition. Together with voice recognition and natural language processing it has enabled the widespread use of digital technologies for personal identification. Our faces, captured by the sensors installed in cameras in public spaces and increasingly in private homes, are translated into data that register our identity. They stand for a person and for a body.

Unlike a passport photograph, a birth certificate or a signature given in a notary's office, digital identities are fluid. They can be connected immediately with everything else known about us and with the networks in which we connect with others. Facial recognition is deployed for many purposes. Cameras are installed for crime prevention and detection, for shortening queues at airports and other identification points, to access buildings or to process financial and other business transactions. We applaud some of these systems and support their aims, yet we feel uneasy about what else might be done with our data. It seems perfectly reasonable for digital technology to monitor and trace the contacts of those infected with the SARS-CoV-2 virus, especially when billions of people around the world face lockdowns. Yet the rate of adoption of tracing apps has remained far below expectations, failing to reach the take-up percentage they need in order to work as they should. And this despite the incorporation of safety standards approved by privacy advocacy groups in line with European Commission regulations, which developers in anticipation of resistance were careful to adopt. The distrust of governments – in particular that emergency measures would persist after the crisis – proved to be too great.

Distrust of AI can resurface at any moment when we feel that our privacy might be threatened, even as we continue to sign away our vaguely defined privacy rights for the sake of digital convenience. Our attitudes towards privacy are sometimes completely contradictory. We cherish it, but are not sure what exactly it entails. We know that we are targeted by advertisements and that much of our data must be stored somewhere. We oscillate between outrage and continuing to upload private content to social media. We have learned to live with privacy erosion, but sometimes we fall into the opposite attitude, wildly defending our rights as though we are members of a tribe for whom to be photographed means to have their soul snatched away. We tend to forget that the notion of privacy is a historically contingent category, one that changes according to the demands of the culture and the economic system in which we happen to live. Some even maintain provocatively that privacy is overestimated, suggesting it is an illusion and cannot be saved, nor needs to be saved (DeBrabander 2020).

Although privacy seems so deeply integral to human existence, it is a fairly recent phenomenon. In a feudal or monarchical society the distinction between public and private made no sense. Take the ceremony of the *lévée* of Louis XIV, the absolute monarch of France. By rendering an intimate scene public, the king at once signalled that private and public were fused in his *persona* and in whatever he did, while demonstrating his power through his accessibility to those selected by him. A contemporary version, or rather, a ludicrous imitation, can be found in Instagram stories. Celebrities and influencers render the seemingly private public, suggesting to their followers that they are in control of both.

The notion of privacy arose in the wake of the ascent of the bourgeoisie in Europe and its success in carving out a public sphere for free debate, protected from censure and intrusion by the state. This began in England, followed by France and Germany in the late eighteenth and nineteenth centuries, where the public sphere was centred on public spaces, such as cafés or pubs, in which discussions among equals could be conducted. These articulations of free speech became public opinion, to be printed in widely circulating newspapers and other news outlets. Buttressed by growing literacy and supported by liberal thinkers, a liberal constitutional order confirmed the separation between the public and the private spheres. This distinction flourished under an early laissez-faire and largely pre-industrial capitalist order, and persisted through the era of industrialization and the rise of mass consumption and the welfare state (Habermas 1962).

The separation between public and private spheres introduced further dichotomies. The public sphere became the space for gainful employment for the sole male breadwinner, while the private home was reserved for bourgeois family life. It was overseen by women, whose appearance in public was strictly regulated by social norms, and functioned as a cherished retreat for the male head of the household. The separation also soon led to a polarization of activities and domains that were marked as male and female and became associated with attributes like rational vs. emotional, objective vs. subjective, strong vs. weak. To this day, such dichotomous thinking feeds gender stereotypes and translates into a persistent pay-gap between male and female workers (Hausen 2012).

Once the power balance between state and market shifted in favour of the latter, privacy lost its connotation as a sphere inviolable to the state's intrusion. It began to shrink in political significance as economic considerations became more important. Shoshana Zuboff writes about living with 'surveillance capitalism', an economic order in which we voluntarily give up our right to privacy in return for the economic benefits we have become conditioned to crave. We are not so much goaded by a well-meaning government or nudged by a clever, far-sighted bureaucracy to divulge information about our location, cultural tastes and individual preferences. Rather, we have developed an intimate form of dependency on giant corporations and the private sector for the amenities, digital gadgets and services they provide us with, while we gladly furnish our data in return. It seems to be the ultimate win–win situation, except that we have to pay a heavy price that many of us are not even aware of (Zuboff 2018).

Once the internet arrived and social media began to dominate every form of communication, a new public sphere was generated in which vested economic interests and political ambitions vie for power, while individuals feel empowered to say whatever they like – even if it is hate speech. But this has not quelled the fear of losing one's privacy, nor has the fear of state surveillance completely vanished. After all, it is still the state, and not the transnational corporation, that retains the monopoly of violence. But what is the price we have to pay for giving up our privacy to a large corporation? At its core, Carissa Véliz argues, privacy matters because the lack of it gives others power over you (Véliz 2020). It entails a loss of individual autonomy when others gain access to what you think and feel. Again, we are reminded of the archaic fear of being possessed by ghosts and demons, or the belief that being photographed means your soul has been taken. Whoever has power over recognizing your face has power over your mind and body.

'Taking back control' has become a political slogan, but it is also at the heart of the continued struggle for autonomy and for a unique sense of self which is in the process of being redefined by advances in the life sciences and in what AI can accomplish. But it is not a zero-sum game. The human mind is adaptable, and how far we will succeed in curbing surveillance depends on the kind of society we live in, which resources we have at our disposal and how determined we are to resist further intrusion. Many well-founded concerns arise from the potential economic and social consequences of having no control over our own data. They relate to living with capitalism, as Zuboff rightly says, and may result in discriminatory practices by insurance companies or employers, or in not having access to housing or other services that used to be communal but are now in the hands of private companies. The predictive power of algorithms and the accessibility of our data for present and future employers, insurance companies, credit-rating firms, housing agencies and other private companies, implies that without strict regulation we will be at their mercy. Regulation can only come from the state, which has to include itself in the regulation. This is rendered even more difficult in a global economic order without a global political order.

The nightmare of state surveillance is no longer linked to Bentham's Panopticon, the utilitarian vision of a perfect society subject to centralized control. For most of us today, such centralization has become inconceivable. We have also forgotten what censorship meant for those who lived under a totalitarian regime in the twentieth century. But something can be learnt from the more subtle and insidious pressures that were at work in the minds of the true believers and followers of totalitarian regimes. In *The Captive Mind*, the Polish Nobel laureate Czesław Miłosz presents a chilling picture of the coping strategies deployed by intellectuals under communism, ranging from fascination to contempt, from aesthetic escape to blind belief – enslavement through consciousness, as he calls it in the final chapter (Miłosz 1953).

Today, inducing citizens to spy on and denounce each other has become obsolete, even in authoritarian regimes. Powerful search engines that rank and prioritize desirable information while excluding and suppressing unwanted information perform much better. Facial recognition of millions of citizens ensures that the control of a population runs efficiently and smoothly through digital means. Despite the concentration of power in a few large transnational corporations, the monopoly of violence is still with the state, with its capacity to turn against its own citizens. This is the basis of dystopian visions deeply rooted in human experience and engraved in the collective memory. They are about power and legitimacy, and about how a political community can become transformed to serve the greed for money and power of those who govern and trample on the common good. The rule of law, human rights and the mutual obligations that regulate the political order of living together must therefore be reaffirmed under the emerging digital order. The issue between citizens and their governments is not so much one of trust, but of trustworthiness – judging their government as being competent, honest and accountable (O'Neill 2018). This now includes what happens to our data.

When, from the mid-eighteenth century onward, statistics became the mode of efficiently governing a growing population under a strengthened state administration, people were classified according to age, gender, height, origin, residency, religion, criminal record and other accessible characteristics. This was the beginning of what Ian Hacking describes as 'Making Up People' – the sorting of individuals into predesigned categories and dimensions considered useful for the purposes of the state bureaucracy, foremost the military, but extending into taxation, public hygiene and mental health (Hacking 1975). The drive for quantification and the trust in numbers in the pursuit of objectivity has defined modern public policy and governance ever since (Porter 1995).

Big data is therefore not as new as is often thought. From a systems theory perspective, digitalization is merely the solution to the problems generated by modern societies. They are organized by functional differentiation and huge amounts of data and information are generated that underpin the regularities of the existing social order. Digitalization thus functions to manage the growing complexity by condensing the self-description of society contained in the data and information it produces, while simultaneously mobilizing energies that are channelled into further societal transformation. The binary code of digitalization enables the discovery of patterns that otherwise would go unobserved, thus updating the knowledge a society has about itself while providing the required stability. Digital technologies merely continue the practices and methods of the social sciences, but aim at a more accurate self-description of the social system by other means (Nassehi 2019).

On the one hand, this functional, systems-theoretical explanation of why digitalization has become a defining and ubiquitous characteristic of contemporary societies is comforting for all those who fear it will destroy their way of life and everything they care for. It tells them that 'governance by numbers' has been practised since the onset of the modern state and that it laid the ground for the modern social order. On the other hand, to rely on a functionalist understanding of how society comes up with new solutions to the problems it creates for itself will be unsatisfactory for those seeking to understand how digitalization works and how it will affect citizens.

It is unlikely to convince them that everything will be done to protect their privacy, especially when they witness the ease and efficiency with which computation-based technologies are applied across different contexts. The sensors used to monitor potholes on roads, sunlight radiation on windowpanes and the functioning of technical equipment in aeroplanes or in hi-tech environments are essentially the same as those used to monitor human bodies. So are the mathematical methods and computational infrastructures that underlie these operations. The software programs deployed to track the movements of terrorists are essentially the same as those tracking the contacts of individuals infected with a virus.

The same holds for the trustworthiness of other privacy-protecting measures. Rare are the moments in history when, given a choice between security and freedom, people chose the latter. Security and surveillance can always be sold as being for our own good, while the idea of freedom remains elusive. The power of prediction that resides in digital technologies, with their ability to jump ahead in time, forces us to rethink the temporal dimension in the option between security and freedom. It may not be so clear-cut.

This existential dilemma is an old one. It reappears in the fascination with machines that can act and decide 'objectively' since they are not subject to human passions and interests. One of the recurring images in politics is a figure from the seventeenth century, Thomas Hobbes's Leviathan. The state is pictured by Hobbes as an automaton, a kind of robot brought to life through artificial motion. It does not think for itself and has no thoughts apart from those given to it by its human component parts. A close look at the famous picture of the Leviathan reveals that it is made of tiny human bodies. It functions by turning human inputs into rational outcomes while stripping humans of their capacity to feed violence and mistrust against each other.

Hobbes's robot is meant to be scary – no individual should think of taking it on – but it is also intended to be reassuring. It was meant to pacify a strife-ridden society that was at war with itself. It offered protection and security at the price of giving up individual freedom. Hobbes's Leviathan was designed to rescue citizens from their natural instinct to fight one another without mercy. Today, the Leviathan would come to our rescue by embodying the ultimate algorithm that calculates what is in the best interest of the state and its citizens and decides for them accordingly. It would protect them from threats coming from inside or outside, from terrorism or from a virus. It would ban the tedious, quarrelsome and unattractive procedures which are part of the functioning of liberal democracies in reaching a compromise. Instead, it would offer short-cuts, or the illusionary solution of re-introducing direct democracy in plural societies.

The Leviathan may change its form and appearance, but its lure is astonishingly enduring. Its lasting appeal is rooted in the desire to overcome political strife and disunity. Endless struggles and squabbles, weak compromises that dissolve shortly after having been reached, pave the way for strong leaders. Most have turned out to be corrupt and greedy, betraying the hopes their followers initially pinned on them. Given the fallibility of human leaders, a better mechanism could replace them, one that is objective and assures impartiality and efficiency and hence is capable of putting an end to the political struggles and grievances.

In Athenian democracy the casting of lots was used to overcome an impasse that could not be resolved otherwise. The Enlightenment thinker Condorcet proposed a rational voting system based on mathematical calculations. Later, economists devised utility functions, and calculating the costs and benefits of trade-offs is standard procedure. The flirtation with an AI as the saving mechanism for governance is only the most recent manifestation of the desire to entrust a formulism or an automaton with the power to solve conflicts and to overcome political strife in an efficient way. Why not go the extra mile and also entrust legislation to an

AI, as we are already habituated to the use of AI in political campaigning? The yearning for a mode of governance that would fit the technological sophistication of the twenty-first century is never far away.

The objections to such latent desires are self-evident. They range from warning against the vulnerability of the technology to hacking and other subversive or malicious abuse, to more fundamental principles. Would a transfer of voting rights or entrusting legislation or the judiciary to an AI not be the end of liberal democracy? Even if minority rights were protected, would it not risk turning political life into a dystopian nightmare, a digitalized dictatorship worse than any previous one? But before we conjure the end of liberal democracy as a result of an AI taking over, let us look at some of the more probable threats that have been discussed recently, some triggered, alas with premonition, by the Trump presidency (Levitsky and Ziblatt 2018).

An inquiry into the risks of what can go wrong in liberal democracies has been undertaken by David Runciman. For him, the end of democracy will not be the result of tanks in the street, riots and the undoing of legitimate institutions by force. These are outdated images from the twentieth century, unlikely to repeat themselves with citizens waking up one morning to discover a coup d'état is taking place (Runciman 2018). And yet – this could almost have happened when the world watched with dismay the storming of the US Capitol by a rioting mob of Trump supporters on 6 January 2021. In comparison to what political violence meant for earlier generations, the sight of the intruders gleefully taking selfies and posturing in bizarre costumes for pictures to be shared on the social media may indicate that Western societies have become more peaceful, but it hardly bodes well for the future of liberal democracies if anti-government sentiment combined with online extremism develops into a violence-prone political movement.

One way in which democracy might come to an end, mentioned by Runciman, is indeed the possibility of a techno-logical takeover, with semi-intelligent machines mining data for us and from us and stealthily taking decisions we are too busy to make. Digital technologies have infiltrated not only our consumption behaviour and social relationships, but also

the functioning of political processes. Political parties everywhere rely on automated databases to run their campaigns.
Trolls and bots steered by invisible hackers, foreign, domestic
or both, can influence public opinion and falsify election
results. Twenty-first-century advertisement has mastered the
art of preying on our cognitive biases, deliberately using them
to make sure we stay on course. With the rise of social media,
the legitimacy of representative democracy is increasingly
questioned, as social media suggest that everyone can and
should be heard and participate directly.

Representative democracy was originally intended to work
against our cognitive biases by putting up barriers against
immediate gratification and filtering the political impulses of
people through institutions designed to correct their biases.
Today, many liberal democracies are witnessing an uneasy
and unresolved competition between the sovereignty of an
elected parliament that represents the will of the people and
the vaguely defined ideal of a direct democracy that allows
people to 'take back control'. As recent examples show, this
crisis serves mainly to empower populists and authoritarians
who claim sovereignty for the nation and for themselves by
seeking to curb a sovereign parliament.

In Runciman's view, Western liberal democracies are
undergoing a mid-life crisis. The crisis is real, but it is too
early to write them off. Maybe he is right and we are in
for a drawn-out demise. If only politics can rescue politics,
meaning that the power of digital technologies has to be
recaptured for democratic politics, then democracy must
be made more responsive than it is right now, and digital
technologies will have to explore new possibilities to find
better solutions for complex issues. The time for intellectual
self-defence has arrived, as Ilija Trojanow, an author and
political essayist, argues. This means sharpening our sense
of self-awareness and guarding against the confirmation bias
that makes us focus only on what reinforces our opinion. It
requires new means of political self-defence, based on the
knowledge of how to deal with information overkill, fake
news, propaganda, trolls and bots so that we do not become
easy prey for whoever wants to manipulate us. In short, it
requires the cultivation of wisdom.

Runciman reports on time he spent in the company of

people whose professional days include worrying about existential risks. These researchers working on intelligent robots, planetary destruction and a host of other risks deemed to be existential were mild-mannered and well-meaning. But to his surprise, they were not interested in politics. Politics for them was a distraction from the overriding question of whether intelligent machines will escape human control. The problems that keep them awake at night were framed in a purely technical way; there was no room for a political dimension. Runciman comments that trying to prevent politics from interfering with the project of saving the world is a noble impulse. But it is a mistake to think that democracy can be contained in a well-tended corner of the garden. It does not take kindly to being contained. When people feel they are being taken for granted they react badly to well-meaning experts taking the important decisions for them. Soon enough, they will reach for whatever grievances they have at hand (Runciman 2018: 111).

In the world today there is no lack of grievances. Many distortions, interferences and bad outcomes are magnified by the use and abuse of digital technologies, but they are hardly caused by them. The realm of politics has always been a fertile ground for metaphors, myths and imaginaries. One recurring and prominent metaphor whose origin precedes industrialization is that of the machine. The metaphor for the modern democratic state was the machine with its self-regulating mechanisms and its checks and balances. The metaphor is still with us, but the machine is no longer made of nuts and bolts working together. The machinery is electronic, consisting of networks and data and driven by predictive algorithms. Now as then, the machine stands for greater efficiency and smooth functioning, but also for a powerful and objective device that may take over from humans. In this imaginary, machines would be accorded the power to replace precarious institutions and practices, in the hope of overcoming the fragility of reaching consensus among the polity. The idea of such a handover to machines is both fascinating and scary.

Some of the earliest foundations for predictive analytics in the political arena were laid by the US firm Simulmatics, identified by the media as the dark force behind John F.

Kennedy's victorious election campaign in 1960. As an ancestor of Google and Facebook, the firm injected big data and computing power into a political context filled with Cold War paranoia, soon followed by the Vietnam War and student protests. Jill Lepore details the firm's mission 'to invent the future' by splitting the US voting population into 480 voter types, and cross-checking with previous voting behaviour to forecast future behaviour. The firm's high ambitions fell flat, and it ended in bankruptcy after a decade. But its core mission still resonates and it became the blueprint for many other profit-seeking firms: 'Collect data. Write code. Detect patterns. Target ads. Predict behaviour. Direct action. Encourage consumption. Influence elections' (Lepore 2020).

The relative ease of political manipulation is one of the reasons why current liberal democracies face what Yaron Ezrahi calls 'a bottomless vacuum in grounding existing power structures'. The influence of modern science on political thought began with the focus on a visual culture in the early modern state, followed by the transfer of the norms of objectivity, representativeness and accountability into politics. Contemporary liberal democracies must be seen as what they are – a necessary fiction. 'A democracy, like any other political regime, must be imagined and performed by multiple agencies in order to exist' (Ezrahi 2012: 1). The necessary fiction shapes the manifold imaginations of what democracy is, what it should look like and how it is performed. In the end, it is up to us to choose the collective imagination of the democracy we want to live by. Fiction extends the range of what is possible. Necessary fictions bridge the gap between our ideals and reality, based on the hope that more of the former will be turned into the latter (Ezrahi 2012).

Hobbes's Leviathan was one of these early fictions in the political imagination. But as a robot it was not the only artificial creature around. Turning the state into a giant automaton incurred the risk that it would not be strong enough. In a world full of machines, it could become entirely artificial and hence indistinguishable from other, perhaps more powerful, machines. Interestingly, the machines that frightened Hobbes most were another species of robot. They were man-made monsters – the corporations of his time, in his view unnatural assemblages of human beings, given artificial life in order to do their bidding. They had no soul or conscience. They could live longer than humans and, like robots, could emerge unscathed from the wreckage of human affairs. Hobbes believed that the only way to control such corporations was to empower the artificial state (Runciman 2018: 128–31).

Before the eighteenth century, states and corporations competed for territory and influence. The most powerful of the corporations, the East India Company, outperformed and outmatched the state in many parts of the world. It was the first global corporate power, collecting taxes and maintaining its own army. It was nationalised by the British state only in the mid-nineteenth century (Dalrymple 2019). At the beginning of the twentieth century, it was Roosevelt – or rather the vast American political machine with Roosevelt as its human face – that succeeded in breaking the monopoly of America's largest corporations. It is sobering to observe that Hobbes's man-made monsters, the corporations of his time, were as dominant then as their successors are now. They no longer trade in spices and arms, tea and other exotic products, but in the extraction of data that is indispensable for the digital oligopolies they now run. In many respects, the state appears weak against their power, especially since corporations ignore national boundaries and have successfully imposed their transnational business models across the world.

Wisdom is certainly needed when reflecting on the regression of liberal democracy, but a twenty-first-century version of a Leviathan run by algorithms is unlikely. Our efforts should rather be directed against human beings usurping power with the help of an AI. We can begin to rethink privacy more comprehensively, with individual autonomy, which is never absolute, at its core. How much are we willing to give up and what do we aspire to receive in return? In Hobbes's version, the individuals making up the Leviathan's component parts give up their freedom for security and protection. Desirable as this must seem in times of civil war and endless violence, the digital age cannot be conceived in such mechanistic terms. It is all about networks and connectivity. If our aim is for a renewed political order, then individual autonomy begins with acknowledging our multiple connectivities and interdependence.

5

Disruption: From BC (Before COVID-19) to AD (After Domestication)

What a biological virus teaches us about our bodies in a digital world

Disruption came, but not as imagined or predicted. It did not come from disruptive technologies, as foretold by management gurus and innovation experts. They had exhorted existing firms to brace themselves for unexpected challenges from unknown and small competitors riding the latest wave of technological advance (Christensen 1997). Disruptive innovation became a universal mantra, feared by some and welcomed by others. It would soon sweep aside everything. But the real disruptor turned out to be a tiny virus, a product of the biological innovation in which evolution is persistently engaged, tinkering with the DNA of various organisms. Its shift from harmless to dangerous occurred when the pathogen switched from an animal to a human host under still unknown circumstances. It is well known that the conditions favourable for zoonosis include the loss of animals' natural habitat and the stressful surroundings under which they are kept for breeding or as food. The virus reminds us that we co-habit this planet with other species and organisms, some of which turn out to be highly aggressive and even lethal for us.

What lessons can be drawn from this tragic period that will continue to preoccupy us in the future? The disruptive virus not only triggered the worst pandemic in a century, but also had the major effect of pushing us faster and further into digitalization (Nowotny 2021). The rest of the story is still unfolding. Even if several effective vaccines have already been dispensed, ensuring continuing vaccination at a global scale will remain challenging. It will take some time to overcome the economic and social harms the virus has caused worldwide. In times of crisis, hidden structures come to the fore. The COVID-19 pandemic mercilessly revealed just how badly prepared most governments were. Health systems turned out to be inadequate to meet the predicted demand for Intensive Care Units, and many ended up in dire straits due to past austerity measures. Ventilators, face masks and protective gear were lacking because their production had been outsourced, exposing the mutual inter-dependence of technology, globalization and crucial human resources. Triage often remained the tragic last option. The 'mathematics of life and death' that underlies the simulation models of contagion rates and infection routes exposed those particularly vulnerable to the disease (Enserink and Kupferschmidt 2020).

But the pandemic revealed more than just unprepared governments, more than the awkward handling of an emergency, and more than the capacity of humans for admirable acts of altruism and solidarity as well as reckless egoism. It also revealed the spooky similarities between a biological virus and a digital virus. Neither is a living organism, but both come deceptively close to it. A virus replicates once it succeeds in infecting a living cell, but it cannot reproduce itself. Its genome can mutate and thus evade the antibodies that make up the vaccines designed to protect against infection. The term 'viral' was used early on to denote the rapid spread of messages through digital media. Like contagion through a biological virus, the diffusion of a message through the internet is intended to 'infect' recipients. The message is inserted either into a piece of DNA or into a text or an image. Both kinds of viruses can wreak havoc on the organism or system they attack, be it genetic or informational. 'Going viral' refers to an accelerating circulation in the system.

Early media theorists, such as Jean Baudrillard and Vilém Flusser, were fascinated by this parallel when trying to understand and follow the effects of novel communication and information media on society. When the carrier is a biological pathogen, the term 'viral' sheds its metaphorical use and reverts to its original, body-threatening meaning. During the COVID-19 pandemic, however, messages about and images of the pathogen were also rapidly distributed. Highly contagious fake news items and dangerous lies were allowed to circulate, inciting people not to wear masks and to refuse vaccination. It was as though the biological virus had been recreated in the mirror world of social media to be as contagious and potentially dangerous as it was in the physical world. 'Viral' morphed from being merely a fascinating metaphor to become a digital agent, acting in the real world where it could cause as much harm. Moreover, just as genetic surveillance was put in place to monitor ongoing mutations, calls for more digital surveillance of dangerous social media content are gathering pace.

The major strategy early on for halting the spread of the virus was imposition of lockdowns, often belatedly. As the virus can only find a host by overcoming physical distance, social interactions were restricted to maintaining that distance, a policy misnamed 'social distancing'. The prescribed range varied from 6 feet to 1–2 metres and from six seconds up to fifteen minutes, all with far-reaching repercussions. Millions of people were obliged to work from home, while kindergartens, schools and universities were closed. Teaching moved online and doctors saw their patients over video calls. No mega-sport events, visits to cultural sights or museums, concerts or religious ceremonies could take place. From one day to the next we were catapulted from the familiar analogue world into a digital no-man's land where human contact was mostly banned and exceptions strictly regulated.

Yet, for all of us, the lifeline during the pandemic turned out to be digital connectivity. Nobody wanted even to imagine what life would have been like without a smartphone, tablet or computer. Our devices kept the worst of the social isolation at bay, allowing us to communicate at a distance. But once physical distancing was in place, we also discovered that too much digital interaction can cause new forms of mental overload. We found ourselves sitting alone in front of a screen, communicating with others also sitting alone in front of their screens. Video conferencing is experienced as being much more exhausting than meetings in person. Our mind is tricked into believing we are together with others while our body feels their absence, causing a dissonance that is experienced as stressful. The synchronicity established by being physically together is broken when one can only see the faces of others on a screen.

The virus has much to teach us about the workings of our bodies and minds and how we relate to each other in the digital world we have created. Digital togetherness removed from the lived bodily experience quickly became an everyday feature, giving us a foretaste of what is to come. Post-pandemic home working is unlikely to mean fewer working hours. Something decisive is giving way at the interface between the analogue world in which our bodies interact with each other and with nature, and the digital world in which physical distance loses the meaning it has in the real world.

The shift of social interaction into the digital world significantly reduced the range of movement of our bodies in the analogue world, giving rise to the hope that the pandemic might be the long-awaited turning point for humanity that would save nature from further destruction. Blue skies appeared above Beijing and other mega-cities that are normally filled with industrial smog. Birdsong reverted to its natural volume as it no longer had to rise above the surrounding noise. In everyday life, people eagerly sought to reconnect with nature. Those who had no garden to tender, cultivated their flowerpots. But the hope of prolonging the truce with an embattled nature soon evaporated. Empty spaces quickly filled up again, even if seeing aeroplanes grounded around the world felt like a parody of children playing musical chairs. The highways were not turned into meadows and the clear skies above the mega-cities again filled with smog. Cruise ships – another heavy contributor to the world's carbon footprint and the pollution of the oceans

– resumed business to promote dream voyages at sea. It became clear that working from home will not be a solution for pressing environmental problems, nor will low-paid jobs disappear. Workers are still needed for food delivery, cleaning and care work, even if robotic automation might eventually also take over even these tasks.

How would we have lived through the pandemic without the current digital technologies at our disposal? The history of previous pandemics reveals the sheer number of human lives lost, with the Black Death in the mid-fourteenth century wiping out almost half of Europe's population. Without science and modern medicine, counter-measures were restricted to the isolation of the infected. But the most significant and lasting effects were the waves of social unrest, rebellion and violence that inevitably followed. So far, we have been spared such major societal disruptions. Paradoxically, although it was the high degree of interconnectivity that rendered our globalized world vulnerable, this also made us more resilient to withstand the worst shocks that could have happened. The informational redundancies that are also built into the various component parts of the connecting networks came to our rescue.

Throughout history repeated outbreaks of cholera have occurred in densely crowded urban quarters where the most basic hygienic facilities and infrastructures were lacking. Perhaps future generations will look back and see the COVID-19 pandemic as the disease of the digital age in its early years. In the long history of epidemics it appears that not every society is afflicted by viruses in the same way. Every society has particular vulnerabilities to the harms caused by a virus, depending on the kind of society it is (Snowden 2019). Seen from this perspective, the COVID-19 pandemic reveals the vulnerabilities that come with a tightly interconnected global world, linked to the largely unregulated pressures exerted on the natural environment.

Pathogens follow existing networks of transportation and communication to create new infection chains. Supply chains are rendered visible when they become interrupted and the scarcity of spare parts suddenly reveals the long distances they have to travel. When supplies of essential medication became exhausted, the reason was that their production had been outsourced to China. Parents who had travelled to a different country to adopt a child found themselves quarantined there. The closure of borders resulted in a sudden loss of seasonal migrant workers in agriculture and meat production. Care provision, especially for the elderly, neared collapse in many Western European countries as it relies heavily on women coming from abroad. Just as the dependence on physical infrastructures became visible, so did the dependence on human labour crossing national borders.

During the nineteenth century, the public response to outbreaks of cholera and related diseases arising from overcrowded and unhealthy living conditions was the resolute introduction of basic hygiene measures and infrastructures. The management of the newly introduced health systems required the state to expand its administrative-bureaucratic capacities. Following social unrest and political struggles, disease-causing bacteria and viruses were eventually brought under control through vaccination and a largely state-run healthcare system, including welfare provisions and policies intended to cover the entire population. In large parts of the world today, these provisions are still lacking.

In the twenty-first century, physical space has been comple-mented by the vastness of digital space, obliging us to build up new capacities for its management. The transhumanists even dream of leaving their bodies and physical space behind and entering effortlessly into a virtual world. It all started at the beginning of the twentieth century when, for the first time in history, the sending of a message became physically decoupled from the messenger. Transmission no longer needed a moving body, whether a horse, a pigeon or a human. Even mechanically moved containers became superfluous. From now on electromagnetic waves would connect sender and receiver using telegraphs and telephones covering ever larger distances. Vast infrastructures of cables and relays, transmitting stations and electricity supplies were put in place and communication from one to many became replaced by communication from many to many.

This meant that the tribal rituals and face-to-face encounters that humans had developed over millennia were adapted, complemented or replaced by new rituals and communi-cation forms. Radio and TV began to reach mass audiences regardless of their geographical location, while social media can now target each individual in highly personalized ways. Viewers were turned from being mere spectators into self-performers. In the famous words of Marshall McLuhan, the media became the message. It became possible to record and replay anything heard or seen, ignoring time-zones and spatial distance. Thus, physical distancing began long before the COVID-19 pandemic, with the crucial difference that replaying a video or reliving a past episode with someone far away was enacted for reasons of necessity or enjoyment. It was voluntary, unlike the physical distancing we are now obliged to practise.

Digitalization also initiated a big jump in the use of symbols and in the level of abstraction. This is significant in evolutionary terms, because it affects the mind as well as the body. The mind becomes free to roam along data highways and engage in playing games in a virtual world. Bodies are transformed into data points making up networks of inter-action that continuously form and change. The interpretation of the movements of these data points representing bodies in space allows precious insights into the complex systems that emerge from networks. The intricate relationship between mind and body – for so long under the spell of a split between them, at least in the West – is being rearranged in ways we do not yet understand.

What we do know is that the digital apps developed to track the movement and location of bodies in order to contain the spread of a virus can also be used for the purposes of monitoring and surveillance. This is one of the ways in which the pandemic has accelerated a process that has been under way for some time, driving us further into digital space. Nobody planned for a virus to push us across the threshold we seem to have passed, but now that it has, old philosophical questions about the intimate relationship between mind and body resurface: Where is the place for the body in a digital world? How can we reintegrate ourselves into a world which is much vaster than the space humans have explored on this earth?

The requirement of physical distancing opened the flood-gates for transferring many activities into digital space in which neither human touch nor other forms of bodily proximity exist. It is as if the wildest dreams of the proponents of digitalization have come true. Working from home? No problem. Food will be delivered at your door and other forms of e-commerce are waiting for your orders. E-banking has been around for some time, and the demise of cash payments has only been accelerated. E-learning offers everything that teachers and students need and helps to cut costs; the realization of its potential has only been delayed by the stubborn belief of academics that face-to-face discussion is still essential for young minds to mature.

Even the cultural sector has come around to embrace digitalization. In the past, it was reluctant, offering only digital titbits. Museums were proud to lure people into standing in long queues for the unique experience of viewing a masterpiece in one of their carefully curated event-exhibitions. Now, just as in other branches of culture such as theatre and music, new digital experiences have to be invented and conveyed to a physically dispersed public, breaking the intimate connection between artist and audience. New forms of creativity will emerge, but we are losing something we treasured without realizing how precious it was.

If we wake up tomorrow in a society that shuns face-to-face contact and has embraced communication via digital bits instead of via bodies and physical proximity, what will it do to us? Once physical distance has been overridden by tele-communication and social contact transformed into tele-interaction we will have become a tele-society, a society that functions at a distance. We will still have bodies, and bodies will still have needs, especially to be cared for. Having a body implies feeling it and its movements in space, in an *umwelt* of the kind that every living organism inhabits. The landscape we are familiar with – a nature that has been transformed through the ages by digging and burning, slashing and clearing, planting and gardening – will undergo the next phase of human intervention. Already now, the total amount of anthropogenic mass – the material output of human activities globally – exceeds all living biomass (Elhacham et al. 2020).

The travel restrictions resulting from the pandemic made virtual travel popular, with webcams installed in wild places. This was reminiscent of the nineteenth-century Romantic movement that flourished while industrialization destroyed what in retrospect became an idyllic version of life in the countryside. Our current eagerness to conserve and protect what can be saved from the past finds its expression in a range of activities. We map the loss of biodiversity, monitor changes in land use, and deplore vanishing mangroves and the increase in desertification. We feel the impending loss of attachment to something we experience through the senses, through smell and touch, sound and vision, while at the same time the ghost in the digital machine drives us further away from these sensations.

This time, the machine is not made of steel and running on wheels, connecting vast stretches of land and driving Western expansion. Instead, the digital machine, like a virus, sends out signals and transfers information that docks on the receptors of our imagination, generating transhumanist fantasies of finally being able to overcome our earth-bound existence and the cumbersome movements of our bodies. This evokes the old dreams of humanity that found expression in different cultures – to fly like a bird or escape like a snake, to remain young forever and to enhance body and mind. These antique myths and transhumanist fantasies can finally unite. At the same time, the body revolts and craves physical contact. Pandemic fatigue sets in as citizens become restless and yearn for interaction in the real world, which is still foremost the social world.

While the disruption we have experienced was not caused by technological innovation, it has nevertheless accelerated the power inherent in digital technologies and will spawn further innovation. Major ruptures in the social fabric have occurred and existing weaknesses have been exposed. The disruption will allow new layers to emerge. It has taught us that we share the world we live in with other species as well as viruses. Existing viruses will not disappear and new ones will soon be around. The more we expand space for us, the more we restrict the natural environments that are home for them. We are engaged in a dual process of co-evolution. On the one hand, we continue to co-evolve with biological viruses; we co-adapt to each other, viruses through mutations and humans by inventing vaccines. On the other hand, our co-evolution with the digital machines invented and designed by us also continues. These processes show further parallels. The contagiousness of digital viruses, just like that of biological viruses, must be subject to surveillance and eventually controlled. But this will have profound implications for individual freedom and how we organize our collective life together.

The merging of digital space and the physical space of everyday life, reinforced through social distancing, brings major changes with it. It has made us acutely aware that we still have bodies with a strong need to relate to other bodies and to nature around us. In the digital space our minds are left free to wander, but we also feel under increased informational and emotional stress. We will have to find a new balance that bridges physical and digital space, just as digital time has to be integrated into social time. This entails a rearrangement of the relation between mind and body, alongside the reordering of our relationship to nature and the natural environment that is under way. We are in the process of adapting to a world markedly different from what it was before COVID-19. The outcome remains to be seen.

The domestication of work

In the future we will look back at the COVID-19 pandemic and divide life as we experienced it individually and collectively into the time before and the time after. A return to normalcy, if it ever existed, is highly unlikely. We see many changes happening around us already and are dimly aware of longer-term consequences in the making that will transcend the much-hoped-for economic recovery. The hiatus between 'before' and 'after' is already visible in different spheres of life, especially in the shift towards working from home, which is unlikely to be reversed. The caption of an article in *The Economist* describing the extent to which work has moved from office to home, and in which sectors, sums it up: 'The era of domestication of work has begun'.

Indeed, perhaps more so than foreseen by the author of the headline, a far-reaching process of domestication is under way. The contemporary domestication of work is not, however, a return to the once self-sufficient economic unit of the *domus*. What began in the remote past with the domestication of wild plants and the breeding of animals for domestic use continues as we adapt our working and living habits to the globally interconnected digital environment. Undoubtedly, this will involve less commuting for millions of employees, while reinforcing the continuous switch between online and offline worlds that is already part of daily life. The merging of the spheres of work and leisure, of time spent at work and family quality time, has been under way for decades, blurring their boundaries. But now an additional dimension has been added. Working from home is remote only in the sense that it is separated geographically from the office, but otherwise it is fully integrated into the multiple and far-reaching networks that characterize the work organization in the twenty-first century.

The domestication of work creates novel temporal and spatial constraints. We become part of a digital time regime in which our waking hours are filled with e-business and e-learning. There will be e-sports, e-conferencing and e-culture, with e-visits to museums or making e-music together. Many more options are on offer, yet the days will still be no longer than twenty-four hours. Nobody knows yet what long-term transformative effects the various post-pandemic recovery and reconstruction efforts will have. But they are unlikely to decrease the high degree of connectivity we have already attained. If the digitalization of society is the answer to the complexity it has reached, as Nassehi asserts, more digitalization is to follow (Nassehi 2019). But how and where, and with which effects on whom?

Long before the pandemic, the future of work was an issue hotly debated by economists, the media and politicians. Although the effects of previous waves of automation on the labour market were reasonably well absorbed, the shared assumption is still that digitalization will lead to aggressive cuts of professional and middle-class jobs. New jobs will certainly be created, but nobody can tell when, nor whether there will remain enough time to soften the blow of mass unemployment. Even an otherwise optimistic and tech-savvy entrepreneur such as Kai-Fu Lee, with ample experience in China and the US, has warned that we should worry more about further economic stratification, inequalities and unemployment in our imminent automated future than speculate about when an Artificial General Intelligence will take over (Lee 2018).

Joseph Schumpeter's insights into the origins and effects of innovation are more than a century old, but they are reconfirmed by analysis of the techno-economic paradigm shifts that follow radical innovation. Schumpeter was one of the few economists who identified innovation, together with entrepreneurship and market power, as a critical dimension of economic change. Innovation, he argued, is a double-edged sword. 'The gale of creative destruction' incessantly undermines the old economic order from within, while creating a new one, changing the fundamentals of how the economy and society work. The destructive part manifests itself in the elimination of skills and industries, devastating entire regions. Economic historians continue to fill out empirically the big picture of creative destruction, showing that, with each shift, sharp income gaps arise between winners and losers and a pervasive mentality of 'winner takes all' takes over. Each of the previous major techno-economic paradigm shifts has engendered a quick concentration of wealth in the hands of a few entrepreneurs and of bold but ruthless investors and speculators. In the end, governments had to step in, to ward off social unrest and/or to pursue a more solidary and progressive political vision (Perez 2018).

New technologies impact the labour market in different ways. The historical patterns that follow the introduction of new technologies have been summarized as a combination of 'bounty' and 'spread'. Bounty stands for the wealth-creating capacity of the new technologies and its concentration, while spread entails an increase in inequalities (Brynjolfsson and McAfee 2014). Future outcomes are again likely to hinge on whether governments can muster the will and political clout to regulate the industries that have become transnational in their operations. But there is nothing inevitable about technology-induced effects on the labour market. When a report written by two academics suggested that 47 per cent of American jobs are at high risk of automation by the mid-2030s, this figure was avidly adopted by the media and politicians alike (Frey and Osborne 2017). It became something of an iconic and unquestioned number, prompting one of the authors to withdraw it as a misunderstanding and offer a more nuanced historical treatise on the social effects of technological change (Frey 2019). Soon, new statistical estimates and scenarios about future job losses will appear, updated in light of the economic and social consequences of the pandemic, and predicting that digitalization will further accelerate the transition.

However, now is the time to bring the core of the real problem into the open: how the meaning of work changes when it is increasingly based on multiple ways of cooperating with digital machines. The concept of work has a long history, reaching back to Aristotle. One strand takes up the ancient dream of liberation from the drudgery and sufferings associated with work. In this version, machines offer relief and emancipation from the kinds of labour better left to animals or machines. It revives the attractive idea of active leisure that led John Maynard Keynes in 1929 to worry about how people would fill their free time when, a century later, the work week had been reduced to eighteen hours (Keynes 1931). Many of the other predictions made in Keynes's speech turned out to be correct, but this one seemed to be far off the mark. We have not yet reached the year 2029, but the arduous recovery of the economy after the COVID-19 meltdown might prove Keynes right after all, though with different worries than his.

Discussion around a Universal Basic Income and its variants is likely to intensify further in the aftermath of the pandemic (Prainsack 2020). It will have to contend with the argument that our economies are driven by wants that are insatiable, and function in ways that mean people need to work. In the words of Robert Skidelsky:

> Supply will always lag behind demand, mandating continuous improvements in efficiency and technology. This will be true even if there is enough to feed, clothe and house the whole world ... humans have no option but to continue to 'work for hire' in whatever jobs the market provides. So the day of abundance, when they can choose between work and leisure, will never arrive. They must race with the machines forever and ever. (Skidelsky 2019)

Such a view echoes the fact that for the largest part of history the economic surplus that went beyond locally prevailing standards of subsistence and comfort was produced by unfree, coerced labour. As late as 1800, up to three-quarters of the world's population was still living in some kind of bondage (Hochschild 2005). Now, we are rapidly approaching a future in which machines are more reliable, efficient and less costly. Regardless of whether or when they may become more intelligent than us, the functioning of the economy and the waging of war may soon no longer need human beings. What will happen once the economic value of each individual tends towards zero? The future of work will depend on how much the value accorded to human beings and their well-being transcends their economic value, and how it will be measured.

A trove of archaeological and archaeo-botanical evidence shows the transition from life in small, mobile, dispersed and relatively egalitarian hunting-and-gathering tribal groups to life in the first walled states that arose in the fertile alluvial crescent between the Euphrates and the Tigris around 3100 BC. This was several millennia after the first crop domestication and after sedentism had occurred, a mixture between hunting, gathering and agricultural practice. In his provocative account of early, agrarian-based state formation in Mesopotamia, James C. Scott argues that it involved the extensive use of unfree labour including war captives, indentured servitude, purchase of slaves from slave markets and forced resettlements of whole communities. The domestication of plants and animals as practised in villages in the near East was extended to domesticate humans.

Grain, so the hypothesis goes, became the ideal means to exert control over the reproduction and labour of the human population that was central for the early states. Grain is ideally suited for spatially concentrated production, tax assessment, appropriation, cadastral surveys, storage and rationing. Grain matures at the same time, is predictable and impossible to hide from the tax collector before the harvest, thus enabling the rise of courtly and administrative elites and early urbanization. Walls were erected as much to prevent the population from fleeing as for the purpose of defence against invaders. The state with the most people was considered the richest, and usually prevailed militarily over smaller rivals. Debt bondage, serfdom, communal bondage and tribute were not confined to Mesopotamia. Slave labour formed the economic basis for Athenian society, imperial Rome and Han Dynasty China. Aristotle held that some people are by nature slaves, due to their lack of rational facilities, and that they were best used, like draft animals, as tools (Scott 2017).

Even if human trafficking and unfree labour continues for millions of people around the world today, the overall contrast is striking. The domestication of humans by humans through forced labour cannot be compared to the gentle domestication facilitated by digitalization of shifting the office into the home. This time no forced labour is involved. We are domesticating ourselves. For the early states the value of the individual labourer was extremely low, but overall numbers mattered and their size was decisive when it came to war, conquest and the expansion of empire. Today, the huge size of the world's still growing population is largely seen as a challenge in terms of striking a balance between the environmental sustainability of the planet and offering access to a decent life and well-being for all. The labour market has become segmented, divided between a highly educated and skilled labour force connected to technology, finance and business and those who find themselves on the other side of the digital divide. We have to prevent it from becoming a digital wall that keeps out those left in the wilderness of a gig economy with low-paid, precarious jobs, unable to flee as there is nowhere left to go.

The issues raised by domesticating ourselves in a digitalized work environment shaped by computers transcend the usual discussions of the future of work. New signals come from computer science, where leading experts like Moshe Vardi have challenged the priority accorded to efficiency in the design and running of computer systems. The relentless pursuit of efficiency, Vardi argues, has decreased the impor-tance given to robustness and resilience. Operations meant to rank relevant search results have become trade secrets that decrease the resilience of the system. Widespread practices such as 'frictionless sharing' end up damaging robustness. A highly promoted and practised 'cyber-libertarianism' that rejects any form of regulation as stifling further innovation has instead generated pervasive cyber-insecurity. Thus, efficiency alone is not optimal. It turns into the 'inefficiency of efficiency' when its focus is exclusively on immediate gains, neglecting resilience which is the only viable and long-term strategy (Vardi 2020).

The future of work remains a hotly discussed topic, as work is still considered to be the lynchpin of our societies in terms of how wealth is allowed to accumulate and how it should be pre- or re-distributed. Arguably, the nature of work is changing more rapidly than other domains affected by ongoing processes of digitalization. The pandemic has only highlighted and accelerated changes that were already under way. Focusing only on the upscaling of digital skills, important as these are for offering people fair opportunities in the labour market, will, however, not be sufficient. The values attributed to different kinds of work, to those who work or no longer work, to those who do not have to work and those who will never be able to, need to be redefined. A wider and deeper discussion of the nature and meaning of work must take place if we are to avoid a society digitally divided between domestic home workers and those left out on the street servicing the other half.

The domestication of work means more than shifting activities from the office to the home. In the evolutionary history of the human species, the domestication of animals and the cultivation of plants arrived together with the practice of agriculture and the establishing of permanent settlements. The first phase of the domestication of work came when freely roaming hunter-gatherers or nomadic pastoralists either switched to farming or were replaced by farmers. As the ways of working to secure a livelihood changed, so did family structures and communal organization, together with a host of cultural practices. Maybe we are now at the threshold of similar far-reaching changes.

From domestication to self-domestication

Only at the onset of the nineteenth century did it become conceivable that humans were domesticating not only animals and plants, adapting them to human use, but also themselves. The concept of self-domestication first gained ground after the discovery of physiological changes in humans that had happened over thousands of years. More evidence emerged for facial-cranial changes such as smaller faces and a shortened snout, the reduction of body mass and sexual dimorphism, and other outward morphological changes. Self-domestication as an evolutionary process is understood as induced self-selection by humans against aggressive behaviour, leading to greater prosociality.

One of the first naturalists to observe such changes in humans was Johann Friedrich Blumenbach. Born half a century before Charles Darwin, he was one of the best-known physiologists and anatomists of his time. As his books were quickly translated into English he reached a wide readership outside his native Germany. Darwin had also intuited the analogy between modern humans and domesticated species, but given his emphasis on controlled breeding, he failed to connect the two phenomena. What was once described as too many ideas chasing too few data has now become a confirmed phenomenon. Thanks to the latest genetic and neurophysiological experimental techniques at the disposal of neurogeneticists, the evidence has accumulated of the shift towards the self-domestication syndrome and has been complemented by an explanatory model for the underlying neural and genetic regulatory mechanisms (Zanella et al. 2019).

The assumption is that the evolutionary changes that drive self-domestication have led to more cooperative and prosocial behaviour. Compared to our closest primate relatives, *Homo sapiens* evolved to be less violent. We are now edging closer to a better understanding of how human aggression has been reduced over the last 300,000 years and why cooperation has been favoured by group-structured cultural selection. One bold, if controversial idea was put forward by Richard Wrangham. His central argument is that the selection of prosocial behaviour and greater docility – sometimes called feminization in the gendered language of science – was due to the evolution of the ability to form coalitions against hyper-aggressive males. This presupposes language and the cognitive ability to plan ahead in premeditated ways. In other words, the subordinates banded together in a conspiracy to kill the physically aggressive alpha male (Wrangham 2019). A faint echo of such behaviour can still be observed in politics today.

Whatever one makes of this explanation – critics quickly pointed out that it neglects the role of female choice in preferring less violent males – it affirms the central importance of understanding what shaped human behaviour in the course of evolutionary history such that humans became more prosocial and enhanced their ability to cooperate. The question of how cooperation arose in societies that increased in size and in the complexity of their organizational structures has also been at the centre of long-standing debates among historians about the evolution of social complexity. One debate revolves around the role played by moralizing gods with their supernatural powers to punish anti-social behaviour. Using proxy indicators from the Seshat database that stores information on more than 450 societies going back as far as 4000 BC, a group of quantitative historians reached the tentative conclusion that the appearance of moralizing gods followed, rather than preceded, a certain level of complexity. It seems that moralizing gods were used as a control instrument to strengthen already-existing power structures, rather than opening the way for centralizing dominance to arise (Whitehouse et al. 2019).

Other historians remain sceptical as they find the quantitative approach far too crude when it comes to understanding what led to the rise of moralizing gods and what the consequences were. They rather see the phenomenon in connection with the genesis of empires, where it is one factor among others. Understandably, they also feel uneasy about what they see as intrusion by outsiders into their disciplinary territory. Nevertheless, cliodynamics, the quantitative approach to history working with large data sets, offers a fertile testing ground for unresolved questions, and more cooperative research will hopefully yield rewarding answers about the rise of socially complex societies.

Long before the arrival of quantitative history, Norbert Elias focused on the social processes that brought about the large-scale changes in cultural attitudes, norms and social behaviour that marked the evolution of Western societies in the post-medieval period. They were characterized by the rise of mutual interdependence and, although Elias never uses the term, an increase in prosocial behaviour. Using data from a large variety of texts, including diaries, manuals of table manners, and forms of speech in addressing people from other social strata, Elias reaches the conclusion that the growing connectivity, the rise of state monopolies (especially over violence and taxing) and the increasing economic interdependencies at a distance led to advances in people identifying with others, irrespective of their social origins.

What started as other-directedness, often by brutal force and with norms imposed from outside, became transformed into self-directedness and self-restraint. New norms and behaviours emerged through what Elias calls the civilizing process. Europeans began to see themselves as more civilized than their ancestors, their neighbours and those they colonized in order to 'civilize' them. The civilizing process unfolded through the links established between domestic and international developments that correlated with changes in the emotional life and the psychological self-perception of individuals. Growing economic and social interdependence required greater stability, coordination and supervision. Elias never tired of insisting that the civilizing process is neither linear nor teleological. He knew from personal experience that a relapse into a de-civilizing barbarism could happen at any time (Elias 1939; Linklater and Mennell 2010).

Taking such evidence from the natural and social sciences, incomplete and contradictory as it is, where are we in the long process of human self-domestication? When assessed on an evolutionary timescale, there has definitely been an increase in the taming of human aggression and a remarkable push towards more cooperation. However, the contemporary picture is murkier. Which historical periods are to be compared and in which regions of the world? The social world is messy, as always. Greater prosociality in some domains contrasts starkly with atrocious aggression and brutality in others. Undoubtedly, we have reached unprecedented levels of social inclusion. New social norms have gained substantial ground as have the demands for social justice and inclusiveness by social movements such as LGBT equality, Black Lives Matter, the Neurodiversity Movement and others. But there is also a growing intolerance and lack of solidarity towards those considered to be outsiders. It is as if social media had removed an inner moral compass, allowing hate speech and contempt to go rampant. Everybody feels entitled, if not incited, to follow their own instincts in defiance of the social order of which they are a part.

Self-domestication as it can be observed at present is not induced by fear of moralizing gods nor imposed by feudal lords. It is not driven by what we associate with Kant's moral imperative, nor by the ideal of *Bildung*, the encompassing educational ideal of self-formation that shaped the ambitious social ascent of the *Bildungsbürgertum* in nineteenth-century Germany. Far from being the presumed universal ideal, this was the privilege of a small Western elite that has long since lost its societal scaffoldings. Inner self-directedness never emerged as a form of auto-genesis. It depended on external social and economic circumstances that enabled cultural values to provide direction, based on the Enlightenment ideal of emancipation.

Gradually, domestication transformed itself into what became a process of self-domestication. It also proceeds when digitalization seeps into our minds and when we rely on algorithms to guide individual behaviour. It affects the space for our bodies in the world and our relationships to each other. Computers and tablets are already daily companions with whom we communicate, exhorting or cursing them, sharing with them our most intimate thoughts and feelings. We trust predictive algorithms and believe they can help us to manage uncertainty. Thus they become part of an extended self. We confer agency on them, just as agency was conferred on moralizing and punishing gods or on other moral figures taken to be setting standards to be obeyed and followed. When big digital corporations align with governments and other businesses in wanting to be seen as seriously engaged in the search for an ethical or responsible AI, their efforts amount to the framing of predictive analytics in ways that align algorithms with the normatively acceptable behaviour they are designed to elicit.

Ever since the first technologies were invented, they have extended the scope and effectiveness of what can be accomplished. With digital technologies programming is a new language that enables algorithms to do much more than calculate. They can adjust, adapt, count, store, control, direct, recognize, compare, amplify, etc. We tell them 'amplify this', 'store that', 'retrieve this', 'sort that' and amazing results follow. It is therefore appropriate, as Arthur suggests, to think of algorithms as verbs (Arthur 2018). Verbs denote actions and algorithms are designed to act and to achieve certain goals. The purpose can be to make money, as in the simple economics of AI. They can be told to map and visualize whatever is found in the surroundings, or be used in apps for finding a restaurant or for surveillance. Self-driving cars can be equipped with evolutionary algorithms, and satellites or drones deployed to monitor worldwide deforestation. Xenobots – tiny organic robots made from two types of stem cells – can navigate inside the body, one of many applications in the rapidly growing area of digitalized medicine and biotechnology. If we think of algorithms as verbs, then we should think of ourselves as nouns. Neither can function on their own. We deploy the technology, but the technology feeds back into our behaviour and we adjust accordingly. This is how self-domestication proceeds.

As always, artists have been at the forefront, making often subversive and creative use of algorithms. In a project named *Machine Auguries*, Alexandra Daisy Ginsberg uses Generative Adversarial Networks to set up a competition between two neural networks. One consists of natural bird song that combines the voices of different kinds of birds, while the other is a Deepfake, a synthetic but indistinguishable artifact that has been trained on data from the voices of the real birds. The Deepfake is then used as playback to the birds so they can enrich the otherwise diminishing variety of their repertoires (Ginsberg 2019).

Just as the birds react in their singing to the synthetic voices they begin to recognize as birds to sing with, so humans react to predictive algorithms that tell them what to expect in the future. The purpose of many algorithms is to make predictions aimed at prevention. They are designed to channel social behaviour in a direction with presumed benefits for the individual and society. Citizens are increasingly held responsible for managing their health and well-being, how they age and how they finance their old age, as provisions once provided by the state are shifted to private companies.

The idea of prevention implies a normative purpose, as prevention aims to avoid harm. In contrast to the precautionary principle, which delays or forestalls action, prevention demands action to avoid harmful consequences in advance, be they road accidents, plastic pollution, obesity or cancer.

In a world perceived to be full of risks and potentially negative fall-out, the scope for prevention is huge. It is another facet of self-domestication. Predictive algorithms want you to live a healthy and productive life, to be happy and to contribute to a sustainable planet. Prevention has become a normative project for the improvement of everything, ideal for the digital age. Besides being beneficial for the individual it saves government spending on public policy measures. It can be a profitable business and making money can easily be combined with the moral satisfaction of being a good citizen. Thus, predictive algorithms that guide us towards appropriate preventive action will assume an ever-greater share of the agency we attribute to them. They become normative agents, geared towards improvement and human enhancement and thus an integral mechanism in the process of self-domestication.

If this is so, we should be careful for what purposes we design them. Self-learning algorithms not only let us see further into the future. They can be trained to bring the future about, in co-production with us. They appear in the guise of technological neutrality and scientific objectivity, which is an added reason why we trust them. The problem remains that the trustworthiness of AI is closely tied to that of the big corporations that own, manipulate and manage them. Power is concentrated in a handful of monopolies that so far have largely escaped regulation. If context matters in how algorithms are deployed, the diversity of contexts has to be acknowledged. This means we have to face what happens when the predictions generated in an artificially restricted context are transferred into the fluid, ambivalent and messy social world in which our future unfolds. We react and adapt to what algorithms predict and it therefore becomes urgent to cultivate critical judgement. Instead of blindly trusting an AI and indiscriminately attributing agency to predictive algorithms, we need to acknowledge the process of self-domestication through digitalization. We domesticate algorithms and in turn they domesticate us. Domesticating them involves taming them. They have to be hauled in from the unregulated wilderness in which they are free to roam and to destroy as they, or rather those who own them, please.

Taming them involves teaching them how to behave, just as we teach children by educating them. We can compare an AI with a child and treat it accordingly. It produces something that often is not very intelligent as it has not yet developed a sense for the social and cultural context in which it operates. Just as a child needs to be familiarized with the culture in which it grows up, an AI needs to learn about the social context in which it will operate. It does not make much sense to criticize a child or an AI for not doing things better if they are not yet sufficiently mature. Educating them means to keep pushing them to develop their potential. Just as we socialize children and instil responsibility in them, we need to teach an AI to acquire the capability to become accountable.

An even bigger challenge for the future is to teach an AI to reflect on why and how things are done, using Judea Pearl's criterion of *I should have acted differently*. To treat algorithms as children we have to take care not to be afraid of them, nor take too seriously what they produce. However, if we neglect to teach them, or train them in the wrong way, they may turn out to be monsters.

At the end, we return to the motto: future needs wisdom. As long as predictive algorithms are not wise – in the sense of having an ethos, a set of shared attitudes and practices attuned to the different contexts in which they are deployed – we have to remain vigilant. Predictive algorithms take us with them on the course they foresee for us. This is the purpose for which they have been designed. But in a messy social world, unforeseen situations arise continuously. They require critical judgement, the agility to act and the right mixture of confidence and humility in embracing uncertainty.

There will always be situations full of ambiguity for which the data extrapolated from the past is insufficient or far too standardized to provide answers relevant to the diversity that pervades local contexts. Algorithmic predictions are extrapolations that presuppose some kind of stability over time in order to extract what is likely to happen in the future. Their power lies in uncovering otherwise hidden patterns and in alerting us to the generation of networks and continuous changes that we are unable to see otherwise. In complex adaptive systems new properties arise and tipping points are reached when the system enters a phase transition and may collapse. Simulations can render such processes visible, allowing us to see further ahead.

But an AI does not know, nor can it predict, a future which remains inherently uncertain and open. It is full of possibilities of which only a tiny fraction will ever be realized. The idea of the future as an open horizon is a great discovery to be treasured. We risk losing it when we start to believe that algorithms can predict the future, especially when it concerns our own behaviour or what will happen to us. This is a powerful message that has to be heeded as we continue to attribute agency to predictive algorithms as part of our co-evolutionary journey with digital machines. If we rely too much on their predictions, we risk returning to a deterministic worldview, where everything has already been decided and where we will find ourselves at the mercy of our own belief in algorithmic predictions.

We do indeed live in exciting times, perhaps another *Sattelzeit*, a temporal watershed in the history of humankind. This time, the entire world is implicated as we find ourselves at the brink of being unable to stave off further degradation of the natural environment, with climate change and other consequences unfolding before our eyes. The increasing overlap between digital space and physical space rearranges the relation between our bodies and minds, as well as how we interact with each other and with what remains of the natural environment. Digital time permeates the other temporalities that we continue to juggle in order to deal with the conflicts between them. On the long evolutionary road of human self-domestication predictive algorithms are now the powerful means through which we domesticate ourselves. They can help us with the many challenges we face. But if we want to retain what it means to be human, we will have to learn to use them wisely and to cultivate the kind of wisdom the future needs.

Bibliography

Agrawal, A., Gans, J. and Goldfarb, A. (2018) *Prediction Machines: The Simple Economics of Artificial Intelligence*, La Vergne: Ingram Publisher Services.

Appiah, A. K. (2018) *The Lies That Bind: Rethinking Identity*, New York: Liveright Publishing.

Arendt, H. (1951) *The Origins of Totalitarianism*, Berlin: Schocken Books.

Arthur, B. (2018) 'How Algorithms are Altering Our Understanding of Systems', in *Exploring Complexity: Volume 7. Grand Challenges for Science in the 21st Century*, https://doi.org/10.1142/11161.

Aubert, M. et al. (2019) 'Earliest Hunting Scene in Prehistoric Art', *Nature* 576, pp. 442–5.

Ball, P. (2019) *How to Grow a Human: Adventures in Who We Are and How We Are Made*, New York: HarperCollins.

Banks, J. (1962) *The Endeavour Journal, 1769–1770*, vol. 1, ed. J. C. Beaglehole, Sydney: Angus and Robertson.

Berlin Debate (2020) Summary, https://www.bosch-stiftung.de/en/project/berlin-science-debate/2020.

Blum, A. (2019) *The Weather Machine: How We See Into the Future*, London: Bodley Head.

Boden, M. (2018) 'Robot Says: Whatever', *Aeon*, 13 August, at https://aeon.co/essays/the-robots-wont-take-over-because-they-couldn't-care-less.

Brynjolfsson, E. and McAfee, A. (2014) *The Second Machine Age: Work, Progress, and Prosperity in a Time of Brilliant Technologies*, New York: W. W. Norton.

Cabanas, E. and Illouz, E. (2019) *Manufacturing Happy Citizens: How the Science and Industry of Happiness Control our Lives*, Cambridge: Polity Press.

Canguilhem, G. (1966) *Le normal et le pathologique*; Paris: Presses Universitaires de France; *The Normal and the Pathological*, trans. C. R. Fawcett, in collaboration with R. S. Cohen, New York: Zone Books, 1991.

Cau, E. (2019) *Il Foglio Innovazione*, 6 August, at https://www.ilfoglio.it/esteri/2019/08/06/news/cosa-fare-con-8chan-il-forum-che-sobilla-gli-attacchi-a-ispanici-e-sinagoghe-268528.

Christensen, C. M. (1997) *The Innovator's Dilemma: When New Technologies Cause Great Firms to Fail*, Boston: Harvard Business School Press.

Cingolani, R. (2019) *L' altra specie. Otto domande su noi e loro*, Bologna: Il Mulino.

Comfort, N. (2019) 'How Science Has Shifted Our Sense of Identity', *Nature* 574, pp. 167–70.

Connor, S. (2019) *The Madness of Knowledge: On Wisdom, Ignorance and Fantasies of Knowing*, London: Reaktion Books.

Cookson, C. (2019) 'A Mini-Revolution in Brain Science', *Financial Times*, 6 September.

Coyle, D. and Weller, A. (2020) '"Explaining" Machine Learning Reveals Policy Challenges', *Science* 368:6498, pp. 1433–4.

Dalrymple, W. (2019) *The Anarchy: The Relentless Rise of the East India Company*, New York: Bloomsbury.

Davies, W. (2015) *The Happiness Industry: How the Government and Big Business Sold Us Well-Being*, New York: Verso.

DeBrabander, F. (2020) *Life after Privacy: Reclaiming Democracy in a Surveillance Society*, Cambridge: Cambridge University Press.

Dehaene, S. (2020) *How We Learn: The New Science of Education and the Brain*, London: Viking.

Delanty, H. (ed.) (2021) *Pandemic, Society and Politics: Critical Reflections on the Pandemic*, Berlin: De Grutyer.

Descola, P. (2019) *Une écologie des relations*, Paris: CNRS Editions.

Dyson, G. (2012) *Turing's Cathedral: The Origins of the Digital Universe*, New York: Pantheon.

Earle, R. (2017) 'Food, Colonialism and the Quantum of Happiness', *History Workshop Journal* 84, pp. 170–93.

Earle, S. (2020) 'Visually Speaking: The New Ubiquity of Photographic Images', *Times Literary Supplement*, 17 January.

Eco, U. (2014) *From the Tree to the Labyrinth: Historical Studies on the Sign and Interpretation*, Cambridge, MA: Harvard University Press.

Economist (2020) 'Citius, Altius, Fortnite: Why the Next Olympics Should Include Fortnite', 27 June, https://www.economist.com/leaders/2020/06/25/why-the-next-olympics-should-include-fortnite.

Elhacham, E. et al. (2020) 'Global Human-Made Mass Exceeds all Living Biomass', *Nature* 588, pp. 442–4.

Elias, N. (1939) *Über den Prozeß der Zivilisation*, Basel: Haus zum Falken.

Elliott, A. (2019) *The Culture of AI: Everyday Life and the Digital Revolution*, Abingdon: Routledge.

Enserink, M. and Kupferschmidt, K. (2020) 'Mathematics of Life and Death: How Disease Models Shape National Shutdowns and Other Pandemic Policies', *Science*, 25 March, doi:10.1126/science.abb8814.

Ernaux, A. (2008) *Les années*, Paris: Gallimard; *The Years*, trans. A. L. Strayer, New York: Seven Stories Press, 2017.

Esposito, E. (2021) *Artificial Communication: How Algorithms Produce Social Intelligence*, manuscript, to be published by MIT Press.

Ezrahi, Y. (2012) *Imagined Democracies: Necessary Political Fictions*, Cambridge: Cambridge University Press.

Felt, U. and Öchsner, S. (2019) 'Reordering the "World of Things": The Sociotechnical Imaginary of RFID Tagging and New Geographies of Responsibility', *Science and Engineering Ethics* 25:5, pp. 1425–46.

Fine, C. (2017) *Testosterone Rex: Myths of Sex, Science, and Society*, New York: W. W. Norton.

Fleurbaey, M. et al. (2018) *A Manifesto for Social Progress: Ideas for a Better Society*, Cambridge: Cambridge University Press.

Ford, P. (2019) 'Why I (Still) Love Tech: In Defense of a Difficult Industry', *WIRED*, 14 May, https://www.wired.com/story/why-we-love-tech-defense-difficult-industry.

Franklin, S. (2019) 'Ethical Research – the Long and Bumpy Road from Shirked to Shared', *Nature* 574, pp. 627–30.

Fraser, J. T. (1975) *Of Time, Passion, and Knowledge: Reflections on the Strategy of Existence*, Princeton: Princeton University Press.

Frey, C. B. (2019) *The Technology Trap: Capital, Labor, and Power in the Age of Automation*, Princeton: Princeton University Press.

Frey, C. B. and Osborne, M. (2017) 'The Future of Employment: How Susceptible are Jobs to Computerisation?', *Technological Forecasting and Social Change* 114, pp. 254–80.

Fry, H. (2018) *Hello World: How to Be Human in the Age of the Machine*, London: Penguin Random House.

Garcia, D. (2017) 'Leaking Privacy and Shadow Profiles in Online

Social Networks', *Science Advances* 3:8, https://doi.org/10.1126/sciadv.1701172.

Ginsberg, A. D. (2019) 'Machine Auguries', https://www.daisyginsberg.com/work/machine-auguries.

Habermas, J. (1962) *Strukturwandel der Öffentlichkeit. Untersuchungen zu einer Kategorie der bürgerlichen Gesellschaft*, Berlin: Suhrkamp; *The Structural Transformation of the Public Sphere: An Inquiry into a Category of Bourgeois Society*, trans. T. Burger and F. Lawrence, Oxford: Blackwell, 1989.

Hacking, I. (1975) *The Emergence of Probability: A Philosophical Study of Early Ideas about Probability, Induction and Statistical Inference*, Cambridge: Cambridge University Press.

Harari, Y. N. (2014) *Sapiens: A Brief History of Humankind*, London: Harvill Secker.

Harari, Y. N. (2018) *21 Lessons for the 21st Century*, London: Jonathan Cape.

Harford, T. (2017) *Fifty Inventions That Shaped the Modern Economy*, New York: Riverhead Books.

Hausen, K. (2012) *Geschlechtergeschichte als Gesellschaftsgeschichte*, Göttingen: Vandenhoeck & Ruprecht.

Hochschild, A. (2005) *Bury the Chains: Prophets and Rebels in the Fight to Free an Empire's Slaves*, Boston: Houghton Mifflin Harcourt.

Holovatch, Y., Kenna, R. and Thurner, S. (2017) 'Complex Systems: Physics Beyond Physics', *European Journal of Physics* 38:2, https://doi.org.10.1088/1361-6404/aa5a87.

Hosni, H. and Vulpiani, A. (2017) 'Forecasting in the Light of Big Data', *Philosophy & Technology* 31, pp. 557–69.

IPBES (2019) *Global Assessment Report on Biodiversity and Ecosystem Services of the Intergovernmental Science-Policy Platform on Biodiversity and Ecosystem Services*, ed. E. S. Brondizio, J. Settele, S. Díaz and H. T. Ngo, Bonn: IPBES secretariat, https://ipbes.net/global-assessment.

Iyer, P. (2019) 'Relative Values: Japan and Authenticity', *Times Literary Supplement*, 22 November.

Jacobson, G. (2019) 'Narratives of Fear', *Times Literary Supplement*, 8 February.

Jasanoff, S. (2007) 'Technologies of Humility, *Nature* 450, p. 33.

Jobin, A., Ienca, M. and Vayena, E. (2019) 'The Global Landscape of AI Ethics Guidelines', *Nature Machine Intelligence* 1, pp. 389–99.

Johnson, S. (2014) *How We Got to Now: Six Innovations That Made the Modern World*, London: Particular Books.

Jorritsma, J. (2020) 'A Future in Ruins: Ghosts, Repetition and the

Presence of the Past in Anthropocenic Futures', *Kronoscope* 20, pp. 190–211.

Kahneman, D. (2011) *Thinking, Fast and Slow*, New York: Macmillan.

Kahneman, D. and Tversky, A. (1979) 'Prospect Theory: An Analysis of Decision under Risk', *Econometrica* 47:2, pp. 263–92.

Kauffman, S. (1995) *At Home in the Universe: The Search for Laws of Self-Organization and Complexity*, Oxford: Oxford University Press.

Keenan, J. P., Gallup, G. G. and Falk, D. (2003) *The Face in the Mirror: The Search of Origins of Consciousness*, New York: Ecco.

Kelly, K. (2019) 'December 2: AR Will Spark the Next Big Tech Platform – Call It Mirrorworld', *WIRED*, 12 February.

Keynes, J. M. (1931) 'Economic Possibilities for our Grandchildren', in *Essays in Persuasion*, London: Macmillan.

King, C. (2019) *Gods of the Upper Air: How a Circle of Renegade Anthropologists Reinvented Race, Sex, and Gender in the Twentieth Century*, New York: Doubleday.

Koselleck, R. (1979) 'Erfahrungsraum und Erwartungshorizont', in *Vergangene Zukunft. Zur Semantik geschichtlicher Zeiten*, Frankfurt: Suhrkamp.

Koselleck, R. (2018) *Sediments of Time: On Possible Histories (Cultural Memory in the Present)*, trans. S. Franzel and S-L. Hoffmann, Palo Alto: Stanford University Press.

Lamont, M. (2000) *The Dignity of Working Men: Morality and the Boundaries of Race, Class, and Immigration*, Cambridge, MA: Harvard University Press; New York: Russell Sage Foundation.

Lamont M. et al. (2016) *Getting Respect: Responding to Stigma and Discrimination in the United States, Brazil, and Israel*, Princeton: Princeton University Press.

Lancaster, M. et al. (2013) 'Cerebral Organoids Model Human Brain Development and Microcephaly', *Nature* 501, pp. 373–9.

Langmead, A. (2019) *Can Computers Do Research?* Presentation at 'What Is Research' Convening, MPIWG, 12–13 June.

Laubichler, M. and Renn, J. (2015) 'Extended Evolution: A Conceptual Framework for Integrating Regulatory Networks and Niche Construction', *Journal of Experimental Zoology Part B: Molecular and Developmental Evolution* 324:7, pp. 565–77.

Le Guin, U. K. (2018) *Dreams Must Explain Themselves: The Selected Non-Fiction of Ursula K. Le Guin*, London: Gollancz.

Lee, E. A. (2020) *The Coevolution: The Entwined Futures of Humans and Machines*, Cambridge, MA: MIT Press.

Lee, K. (2018) *AI Super-Powers: China, Silicon Valley, and the New World Order*, Boston: Houghton Mifflin Harcourt.

Leonelli, S. and Tempini, N. eds. (2020) *Data Journeys in the Sciences*, Cham: Springer Open.

Lepore, J. (2018) 'What 2018 Looked Like Fifty Years Ago', *The New Yorker*, 31 December.

Lepore, J. (2020) *If Then: How the Simulmatics Corporation Invented the Future*, New York: Liveright Publishing Corporation.

Levitsky, S. and Ziblatt, D. (2018) *How Democracies Die*, New York: Crown.

Linklater, A. and Mennell, S. (2010) 'Norbert Elias, *The Civilizing Process: Sociogenetic and Psychogenetic Investigations* – an Overview and Assessment', *History and Theory* 49, pp. 384–411.

Loreto, V. (2018) 'Need a New Idea? Start at the Edge of What is Known', Ted Talk, https://www.ted.com/talks/vittorio_loreto_need_a_new_idea_start_at_the_edge_of_what_is_known?language=enTalk.

Lubbe, A. S. (2019) Personal email correspondence, 16 July.

McAnany, P. et al. (2009) *Questioning Collapse Human Resilience, Ecological Vulnerability, and the Aftermath of Empire*, Cambridge: Cambridge University Press.

McEwan, I. (2018) 'Düssel...', *The New York Review of Books*, 19 July, pp. 4–5.

McNeill, J. R. and Engelke, P. (2015) *The Great Acceleration: An Environmental History of the Anthropocene since 1945*, Cambridge, MA: Harvard University Press.

Maimon-Mor, R. et al. (2017) 'Peri-hand Space Representation in the Absence of a Hand – Evidence from Congenital One-handers', *Cortex* 95, pp. 169–71.

Marshall, M. (2020) *The Genesis Quest: The Geniuses and Eccentrics on a Journey to Uncover the Origin of Life on Earth*, Chicago: University of Chicago Press.

Maynard Smith, J. and Szathmáry, E. (1995) *The Major Transitions in Evolution*, Oxford: Oxford University Press.

Miłosz, C. (1953) *The Captive Mind*, London: Secker and Warburg.

Mokyr, J. (2010) *The Enlightened Economy: Britain and the Industrial Revolution, 1700–1850*, New Haven: Yale University Press.

Mokyr, J. (2016) *A Culture of Growth: The Origins of the Modern Economy* (Graz Schumpeter Lectures), Princeton: Princeton University Press.

Nanetti, A. (2021) 'Defining Heritage Science: A Consilience Pathway to Treasuring the Complexity of Inheritable Human Experiences through Historical Method, AI and ML', *Complexity*, special issue: *Tales of Two Societies: On the Complexity of the Coevolution between the Physical Space and the Cyber Space*, https://doi.org/10.1155/2021/4703820.

Nassehi, A. (2019) *Muster: Theorie der digitalen Gesellschaft*, Munich: C. H. Beck.

Nowotny, H. (1989) *Eigenzeit. Entstehung und Strukturierung eines Zeitgefühls*, Frankfurt: Suhrkamp; *Time: The Modern and Postmodern Experience*, trans. Neville Plaice, Cambridge: Polity Press, 1994.

Nowotny, H. (2017) *An Orderly Mess*, Budapest: CEU Press.

Nowotny, H. (2021) 'In Artificial Intelligence We Trust: How the COVID-19 Pandemic Pushes Us Deeper into Digitalization', in H. Delanty (ed.), *Pandemic, Society and Politics: Critical Reflections on the Pandemic*, Berlin: De Grutyer, 2021.

Nowotny, H. and Schot, J. (2018) 'It Could Be Otherwise: Social Progress, Technology and the Social Sciences', *Technology's Stories* 6:2, https://doi.org/10.15763/jou.ts.2018.05.14.05.

O'Neill, O. (2018) 'Linking Trust to Trustworthiness', *International Journal of Philosophical Studies* 26:2, https://doi.org/10.1080/09 672559.2018.1454637.

Pearl, J. and Mackenzie, D. (2018) *The Book of Why: The New Science of Cause and Effect*, London: Allen Lane.

Perez, C. (2018) 'Second Machine Age or Fifth Technological Revolution? (Part 4) The Historical Patterns of Bounty and Spread', *UCL Institute for Innovation and Public Purpose Blog*, 21 November, http://beyondthetechrevolution.com/blog/ second-machine-age-or-fifth-technological-revolution-part-4.

Perrow, C. (1984) *Normal Accidents: Living with High-risk Technologies*, New York: Basic Books.

Pigliucci, M. and Müller, G. B. (2010) *Evolution: The Extended Synthesis*, Cambridge, MA: MIT Press.

Pinker, S. (2018) *Enlightenment Now: The Case for Reason, Science, Humanism, and Progress*, London: Viking.

Pistoletto, M. (2016) *Hominitheism and Demopraxy*, http://www. pistoletto.it/eng/testi/hominitheism-and-demopraxy.pdf

Pollan, M. (2018) *How to Change Your Mind: The New Science of Psychedelics*, London: Penguin.

Porter, T. M. (1995) *Trust in Numbers: The Pursuit of Objectivity in Science and Public Life*, Princeton: Princeton University Press.

Prainsack, B. (2020) *Vom Wert des Menschen. Warum wir ein bedingungsloses Grundeinkommen brauchen*, Vienna: Brandstätter.

Prajda, K. (2016) 'Manetto di Jacopo Amannatini, the Fat Woodcarver', *Acta Historiae Artium Academiae Scientiarum Hungaricae* 57:1, pp. 5–21.

Preston, E. (2018) 'A "Self-Aware" Fish Raises Doubts About a Cognitive Test', *Quantamagazine*, 12 December, https://www.quantamagazine.org/a-self-aware-fish-raises-doubts-about-a-cognitive-test-20181212.

Reich, D. (2018) *Who We Are And How We Got Here: Ancient DNA and the New Science of the Human Past*, Oxford: Oxford University Press.

Renn, J. (2020) *The Evolution of Knowledge: Rethinking Science for the Anthropocene*, Princeton: Princeton University Press.

Reyes, M. (2015) 'The Ex Machina Ending Debate: Is The Movie 3 Minutes Too Long?', https://www.cinemablend.com/new/Ex-Machina-Ending-Debate-Movie-3-Minutes-Too-Long-71101.html.

Rosling Rönnlund, A., Rosling, O. and Rosling, H. (2018) *Factfulness: Ten Reasons We're Wrong About The World – And Why Things Are Better Than You Think*, London: Sceptre.

Rosol, C. et al. (2018) 'On the Age of Computation in the Epoch of Humankind', https://www.nature.com/articles/d42473-018-00286-8.

Runciman, D. (2018) *How Democracy Ends*, London: Profile Books.

Russell, S. (2019) *Human Compatible: AI and the Problem of Control*, London: Allen Lane.

Scheidel, W. (2019) *Escape from Rome: The Failure of Empire and the Road to Prosperity*, Princeton: Princeton University Press.

Schilthuizen, M. (2018) *Darwin Comes to Town: How the Urban Jungle Drives Evolution*, London: Picador.

Schwartz, H. (1996) *The Culture of the Copy: Striking Likenesses, Unreasonable Facsimiles*, New York: Zone Books.

Scoones, I. and Stirling, A. (2020) *The Politics of Uncertainty: Challenges of Transformation*, Abingdon: Routledge.

Scott, J. (1999) *Seeing Like a State: How Certain Schemes to Improve the Human Condition Have Failed*, New Haven: Yale University Press.

Scott, J. C. (2017) *Against the Grain: A Deep History of the Earliest States*, New Haven: Yale University Press.

Shiller, R. J. (2019) *Narrative Economics: How Stories Go Viral and Drive Major Economic Events*, Princeton: Princeton University Press.

Skidelsky, R. (2019) Twitter post, 15 May, @RSkidelsky.

Slezak, M. (2015) 'Key Moments in Human Evolution Were Shaped by Changing Climate', *New Scientist*, 16 September.

Snowden, F. (2019) *Epidemics and Society: From the Black Death to the Present*, New Haven: Yale University Press.

Sperber, D. (1996) *Explaining Culture: A Naturalistic Approach*, New Jersey: Wiley.

Spicer, A. (2020) 'Sci-fi Author William Gibson: How "Future Fatigue" is Putting People Off the 22nd Century', *The Conversation*, 23 January, https://theconversation.com/sci-fi-author-william-gibson-how-future-fatigue-is-putting-people-off-the-22nd-century-130335.

Steffen, W. et al. (2015) 'The Trajectory of the Anthropocene: The Great Acceleration', *The Anthropocene Review* 2:1, pp. 81–98.

Stokstad, E. (2019) 'The New Potato: Breeders Seek a Breakthrough to Help Farmers Facing an Uncertain Future', *Science* 363, pp. 574–7.

Tainter, J. (1988) *The Collapse of Complex Societies*, Cambridge: Cambridge University Press.

Tegmark, M. (2017) *Life 3.0: Being Human in the Age of Artificial Intelligence*, New York: Vintage.

Tenner, E. (1997) *Why Things Bite Back: Technology and the Revenge of Unintended Consequences*, New York: Vintage.

Thornhill, J. (2018) 'Asia Has Learnt to Love Robots – the West Should, Too', *Financial Times*, 31 May.

Vardi, M. Y. (2020) 'Lessons for Digital Humanism from Covid-19', https://www.youtube.com/watch?v=AIOnixSBAvI.

Véliz, C. (2020) *Privacy is Power: Why and How You Should Take Back Control of Your Data*, London: Bantam Press.

Von Neumann, J. (1955) 'Can We Survive Technology?', http://fortune.com/2013/01/13/can-we-survive-technology.

Von Uexküll, J. (1909) *Umwelt und Innenleben der Tiere*, Berlin: Springer.

Wajcman, J. (2019) 'The Digital Architecture of Time Management', *Science, Technology & Human Values* 44:2, pp. 315–37.

Whitehouse, H. et al. (2019) 'Complex Societies Precede Moralizing Gods Throughout World History', *Nature* 568, pp. 226–9.

Wiener, N. (1960) 'Some Moral and Technical Consequences of Automation', *Science* 131:3410, pp. 1355–8.

Wigderson, A. (2019) *Mathematics and Computation: A Theory Revolutionizing Technology and Science*, Princeton: Princeton University Press.

Winterson, J. (2019) *Frankissstein: A Love Story*, New York: Grove Atlantic.

Wittgenstein, L. (1922) *Tractatus Logico-Philosophicus*, trans. F. P. Ramsey and C. K. Ogden London: Kegan Paul.

Wrangham, R. (2019) *The Goodness Paradox: The Strange Relationship Between Virtue and Violence in Human Evolution*, New York: Pantheon.

Zanella, M. et al. (2019) 'Dosage Analysis of the 7q11.23 Williams Region Identifies BAZ1B as a Major Human Gene Patterning the Modern Human Face and Underlying Self-Domestication', *Science Advances* 5:12, eaaw7908.

Zuboff, S. (2018) *The Age of Surveillance Capitalism: The Fight for a Human Future at the New Frontier of Power*, New York: Hachette Book Group.

Zurn, P. et al. (2020) *Curiosity Studies: Toward A New Ecology of Knowledge*, Minneapolis: University of Minnesota Press.

Index